volume 62

cture notes in pure and applied mathematics

ordered groups

Edited by

Jo E. Smith
G. Otis Kenny
Richard N. Ball

ORDERED GROUPS

Contributions to *Lecture Notes in Pure and Applied Mathematics* are reproduced by direct photography of the author's typewritten manuscript. Potential authors are advised to submit preliminary manuscripts for review purposes. After acceptance, the author is responsible for preparing the final manuscript in camera-ready form, suitable for direct reproduction. Marcel Dekker, Inc. will furnish instructions to authors and special typing paper. Sample pages are reviewed and returned with our suggestions to assure quality control and the most attractive rendering of your manuscript. The publisher will also be happy to supervise and assist in all stages of the preparation of your camera-ready manuscript.

LECTURE NOTES
IN PURE AND APPLIED MATHEMATICS

1. *N. Jacobson,* Exceptional Lie Algebras
2. *L.-Å. Lindahl and F. Poulsen,* Thin Sets in Harmonic Analysis
3. *I. Satake,* Classification Theory of Semi-Simple Algebraic Groups
4. *F. Hirzebruch, W. D. Newmann, and S. S. Koh,* Differentiable Manifolds and Quadratic Forms
5. *I. Chavel,* Riemannian Symmetric Spaces of Rank One
6. *R. B. Burckel,* Characterization of C(X) Among Its Subalgebras
7. *B. R. McDonald, A. R. Magid, and K. C. Smith,* Ring Theory: Proceedings of the Oklahoma Conference
8. *Y.-T. Siu,* Techniques of Extension of Analytic Objects
9. *S. R. Caradus, W. E. Pfaffenberger, and B. Yood,* Calkin Algebras and Algebras of Operators on Banach Spaces
10. *E. O. Roxin, P.-T. Liu, and R. L. Sternberg,* Differential Games and Control Theory
11. *M. Orzech and C. Small,* The Brauer Group of Commutative Rings
12. *S. Thomeier,* Topology and Its Applications
13. *J. M. López and K. A. Ross,* Sidon Sets
14. *W. W. Comfort and S. Negrepontis,* Continuous Pseudometrics
15. *K. McKennon and J. M. Robertson,* Locally Convex Spaces
16. *M. Carmeli and S. Malin,* Representations of the Rotation and Lorentz Groups: An Introduction
17. *G. B. Seligman,* Rational Methods in Lie Algebras
18. *D. G. de Figueiredo,* Functional Analysis: Proceedings of the Brazilian Mathematical Society Symposium
19. *L. Cesari, R. Kannan, and J. D. Schuur,* Nonlinear Functional Analysis and Differential Equations: Proceedings of the Michigan State University Conference
20. *J. J. Schäffer,* Geometry of Spheres in Normed Spaces
21. *K. Yano and M. Kon,* Anti-Invariant Submanifolds
22. *W. V. Vasconcelos,* The Rings of Dimension Two
23. *R. E. Chandler,* Hausdorff Compactifications
24. *S. P. Franklin and B. V. S. Thomas,* Topology: Proceedings of the Memphis State University Conference
25. *S. K. Jain,* Ring Theory: Proceedings of the Ohio University Conference
26. *B. R. McDonald and R. A. Morris,* Ring Theory II: Proceedings of the Second Oklahoma Conference
27. *R. B. Mura and A. Rhemtulla,* Orderable Groups
28. *J. R. Graef,* Stability of Dynamical Systems: Theory and Applications
29. *H.-C. Wang,* Homogeneous Banach Algebras
30. *E. O. Roxin, P.-T. Liu, and R. L. Sternberg,* Differential Games and Control Theory II
31. *R. D. Porter,* Introduction to Fibre Bundles
32. *M. Altman,* Contractors and Contractor Directions Theory and Applications
33. *J. S. Golan,* Decomposition and Dimension in Module Categories
34. *G. Fairweather,* Finite Element Galerkin Methods for Differential Equations
35. *J. D. Sally,* Numbers of Generators of Ideals in Local Rings
36. *S. S. Miller,* Complex Analysis: Proceedings of the S.U.N.Y. Brockport Conference
37. *R. Gordon,* Representation Theory of Algebras: Proceedings of the Philadelphia Conference
38. *M. Goto and F. D. Grosshans,* Semisimple Lie Algebras
39. *A. I. Arruda, N. C. A. da Costa, and R. Chuaqui,* Mathematical Logic: Proceedings of the First Brazilian Conference
40. *F. Van Oystaeyen,* Ring Theory: Proceedings of the 1977 Antwerp Conference

41. *F. Van Oystaeyen and A. Verschoren,* Reflectors and Localization: Application to Sheaf Theory
42. *M. Satyanarayana,* Positively Ordered Semigroups
43. *D. L. Russell,* Mathematics of Finite-Dimensional Control Systems
44. *P.-T. Liu and E. Roxin,* Differential Games and Control Theory III: Proceedings of the Third Kingston Conference, Part A
45. *A. Geramita and J. Seberry,* Orthogonal Designs: Quadratic Forms and Hadamard Matrices
46. *J. Cigler, V. Losert, and P. Michor,* Banach Modules and Functors on Categories of Banach Spaces
47. *P.-T. Liu and J. G. Sutinen,* Control Theory in Mathematical Economics: Proceedings of the Third Kingston Conference, Part B
48. *C. Byrnes,* Partial Differential Equations and Geometry
49. *G. Klambauer,* Problems and Propositions in Analysis
50. *J. Knopfmacher,* Analytic Arithmetic of Algebraic Function Fields
51. *F. Van Oystaeyen,* Ring Theory: Proceedings of the 1978 Antwerp Conference
52. *B. Kedem,* Binary Time Series
53. *J. Barros-Neto and R. A. Artino,* Hypoelliptic Boundary-Value Problems
54. *R. L. Sternberg, A. J. Kalinowski, and J. S. Papadakis,* Nonlinear Partial Differential Equations in Engineering and Applied Science
55. *B. R. McDonald,* Ring Theory and Algebra III: Proceedings of The Third Oklahoma Conference
56. *J. S. Golan,* Structure Sheaves over a Noncommutative Ring
57. *T. V. Narayana, J. G. Williams, and R. M. Mathsen,* Combinatorics, Representation Theory and Statistical Methods in Groups: YOUNG DAY Proceedings
58. *T. A. Burton,* Modeling and Differential Equations in Biology
59. *K. H. Kim and F. W. Roush,* Introduction to Mathematical Consensus Theory
60. *J. Banas and K. Goebel,* Measures of Noncompactness in Banach Spaces
61. *O. A. Nielsen,* Direct Integral Theory
62. *J. E. Smith, G. O. Kenny, and R. N. Ball,* Ordered Groups: Proceedings of the Boise State Conference

Other Volumes in Preparation

ORDERED GROUPS

PROCEEDINGS OF THE BOISE STATE CONFERENCE

edited by

Jo E. Smith
Department of Science and Mathematics
General Motors Institute
Flint, Michigan

G. Otis Kenny
Richard N. Ball
Department of Mathematics
Boise State University
Boise, Idaho

MARCEL DEKKER, INC. New York and Basel

Library of Congress Cataloging in Publication Data

Conference on Ordered Groups, Boise, Idaho, 1978.
Ordered groups.

(Lecture notes in pure and applied mathematics ; 62)
1. Ordered groups--Congresses. I. Smith, Jo E. II. Kenny, G. Otis. III. Ball, Richard N. IV. Boise State University. V. Title.
QA171.C6794 1978 512'.2 80-24251
ISBN 0-8247-6943-0

MARCEL DEKKER, INC.
270 Madison Avenue, New York, New York 10016

Current printing (last digit):

10 9 8 7 6 5 4 3 2 1

PRINTED IN THE UNITED STATES OF AMERICA

This volume is dedicated to the memory of

Boris Zaharovich Vulih
(1913-1978)

whose research in ordered structures formed an integral part of the groundwork on which much of our current work is based.

PREFACE

The papers contained in this volume constitute the proceedings of the Conference on Ordered Groups, which was held in Boise, Idaho from October 16, 1978 to October 20, 1978. This conference was an effort to bring together many of the mathematicians currently working in the field of ordered structures, to share not only the technical results of their research, but also the new uses of ideas and techniques from other areas of mathematics being made in the study of ordered structures. The resulting collection of papers exhibits both the diversity and the depth of current research and serves as a source of problems and ideas for possible subsequent work in this field.

We would like to express our gratitude to Dr. William P. Mech and to the entire faculty of the Department of Mathematics of Boise State University for their encouragement, help, support, and understanding during the months of preparation for and the week of the conference. We also thank the Associated Students of Boise State University who, as student sponsors of the conference, provided the meeting room in which the papers were presented. We would like to acknowledge the partial funding of the conference by the Center for Research, Grants, and Contracts, the Department of Mathematics, and Dean William Keppler, all of Boise State University. The contributions of Mrs. Dianne Ellis toward the production of the conference and the typing of this volume are also greatly appreciated, as is the hard work of our other typists, Arlene DeHaas, Kathy Gordon, Becky Reed and Joan Winn. Lastly, our sincere thanks go to the conference participants and contributors, who made the conference such a rewarding experience for us.

Jo E. Smith
G. Otis Kenny
Richard N. Ball

CONTENTS

CONTRIBUTORS

RICHARD N. BALL
Department of Mathematics, Boise State University, Boise, Idaho

MAUREEN A. BARDWELL
Division of Natural Sciences, St. Norbert College, DePere, Wisconsin

PAUL CONRAD
Department of Mathematics, University of Kansas, Lawrence, Kansas

A. M. W. GLASS
Department of Mathematics and Statistics, Bowling Green State University, Bowling Green, Ohio

YURI GUREVICH[1]
Department of Mathematics, Ben Gurion University of the Negev, Be'er Sheva, Israel

W. CHARLES HOLLAND
Department of Mathematics and Statistics, Bowling Green State University, Bowling Green, Ohio

D. C. KENT
Department of Pure and Applied Mathematics, Washington State University, Pullman, Washington

A. V. KOLDUNOV
Hertzen Pedagogical Institute, Leningrad, U.S.S.R.

STEPHEN H. McCLEARY
Department of Mathematics, University of Georgia, Athens, Georgia

GEORGE F. McNULTY
Department of Mathematics and Computer Science, University of South Carolina, Columbia, South Carolina

JORGE MARTINEZ
Department of Mathematics, University of Florida, Gainesville, Florida

KEITH R. PIERCE
Department of Mathematics, University of Missouri-Columbia, Columbia, Missouri

G. Ya. ROTKOVICH
Hertzen Pedagogical Institute, Leningrad, U.S.S.R.

JO E. SMITH[2]
Department of Mathematics, Boise State University, Boise, Idaho

ELLIOT C. WEINBERG
Department of Mathematics, University of Illinois at Urbana-Champaign, Urbana, Illinois

ROBERT ROSS WILSON
Department of Mathematics, California State University, Long Beach, California

---

[1]Paper presented while on leave to Department of Mathematics, Simon Fraser University, Burnaby, British Columbia

[2]Current address: Department of Science and Mathematics, General Motors Institute, Flint, Michigan

ORDERED GROUPS

THE STRUCTURE OF AN ℓ-GROUP THAT IS DETERMINED BY ITS MINIMAL PRIME SUBGROUPS

Paul Conrad

University of Kansas
Lawrence, Kansas

1. INTRODUCTION

Throughout let G be a lattice-ordered group (ℓ-group) and let $\mathcal{C}(G)$ be the complete distributive lattice of all the convex ℓ-subgroups of G. Then $C \in \mathcal{C}(G)$ is *regular* if it is maximal without some $g \in G$. In this case C is a *value* of g and C and g are called *special* if C is the only value of g. Each convex ℓ-subgroup is the intersection of the regular subgroups that contain it, and each regular subgroup C is *prime* (i.e., $a \wedge b = 0$ implies $a \in C$ or $b \in C$). The set $\{G_\gamma | \gamma \in \Gamma\}$ of all regular subgroups of G is a *root system* (i.e., a po set such that the elements $\supseteq G_\gamma$ form a chain). For these and other results about Γ see [8] or [13] or [3].

For each $\gamma \in \Gamma$ let G^γ be the convex ℓ-subgroup of G that covers G_γ and let $S_\gamma = \{g \in G | \text{no value of } g \text{ is comparable with } \gamma\}$. Then $S_\gamma \in \mathcal{C}(G)$ and this paper is essentially a study of these subgroups S_γ and how they influence the structure of G. If $A \in \mathcal{C}(G)$ then $A' = \{g \in G| \; |g| \wedge |a| = 0$ for all $a \in A\}$ is the *polar* of A. These polars form a complete Boolean algebra [16].

We show that $A' = \cap\{S_\alpha | G_\alpha \not\supseteq A\}$ and

$$\begin{aligned} S_\gamma &= \cap \text{ all the minimal primes contained in } G_\gamma \\ &= \cup \{a' | 0 < a \in G \setminus G_\gamma\} \\ &= \cap \{G_\delta | \delta \text{ and } \gamma \text{ are comparable}\}. \end{aligned}$$

The ℓ-group G is *representable* as a subdirect product of totally ordered groups (o-groups) iff each $S_\gamma \lhd G$. G is *hyperarchimedian* (i.e., each ℓ-homomorphic image of G is archimedian) iff each $S_\gamma = G_\gamma$. G_γ is special iff

$G^\gamma = S_\gamma \boxplus T$ where $\boxplus$ denotes the cardinal sum, and so G is *finite valued* (i.e., each element of G has only a finite number of values) iff $G^\gamma = S_\gamma \boxplus T_\gamma$ for each $\gamma \in \Gamma$.

We say that G is a *lex-group* if it is a lexicographic extension of an ℓ-group by a non-zero o-group, and show that this is the case iff $S_\gamma = 0$ for some $\gamma \in \Gamma$. Moreover, if C is an ℓ-ideal of G then G/C is a lex-group iff $G_\gamma \supseteq C \supseteq S_\gamma$ for some $\gamma \in \Gamma$.

In section 3 we investigate the tree $\mathcal{S} = \{S_\gamma | \gamma \in \Gamma\}$. S_γ and S_β are incomparable (which we shall denote by $S_\gamma || S_\beta$) iff $\gamma || \beta$ in Γ, and $S_\alpha = S_\beta$ iff α and β lie on the same roots in Γ. The following are equivalent: γ is maximal in $\mathcal{S}$; S_γ is prime; S_γ is a minimal prime; G_γ contains a unique minimal prime. In particular $\mathcal{S}$ consists of atoms iff each prime in G exceeds a unique minimal prime. Each minimal prime subgroup of G is a polar iff $\cap_\Lambda S'_\gamma \neq 0$ for each root Λ of Γ.

THEOREM 3.6. For an ℓ-group G the following are equivalent.

(1) There exists a maximal trivially ordered subset $\alpha_1, \ldots, \alpha_n$ of Γ.

(2) There exists a finite subset of $\mathcal{S}$ with zero intersection.

(3) There exists a finite set of special elements, which is also a maximal disjoint subset of G.

(4) Each pairwise disjoint set of roots in Γ is finite.

Then we show that G is a lex-sum of ℓ-groups $a_1'', \ldots, a_n''$ where each a_i is special iff G satisfies (2). In particular, G is a lex-sum of a finite number of o-groups iff $\mathcal{S}$ is finite (this is the so-called *finite basis theorem* for ℓ-groups).

If G is representable and $S_{\gamma_1}, \ldots, S_{\gamma_n}$ is a trivially ordered set of maximal element in $\mathcal{S}$ then $G/\cap S_{\gamma_i}$ has a finite basis of n-elements.

THEOREM 3.10. The set of elements in $\mathcal{S}$ that are cardinal summands of G has zero intersection iff there is an ℓ-isomorphism σ of G such that

$$\Sigma a_\lambda'' \subseteq G\sigma \subseteq \Pi a_\lambda''$$

where the a_λ are special elements of G. If this is the case then the a_λ'' are all of the indecomposable summands of G and the a_λ' are all of the summands of G in $\mathcal{S}$. In particular $G = \Sigma a_\lambda''$ iff $\mathcal{S}$ is atomic and each atom is a summand, and in this case the atoms of $\mathcal{S}$ are the a_λ'. Finally a_λ'' is an o-group iff a_λ' is maximal in $\mathcal{S}$.

This theorem has many interesting corollaries.

In section 4 we investigate an equivalence relation ~ on Γ defined by $\alpha \sim \beta$ if $S_\alpha = S_\beta$. Each equivalence class $\overline{\alpha}$ is a convex chain and $\overline{\Gamma}$ is a root system. For each $\overline{\alpha} \in \overline{\Gamma}$ we have the convex ℓ-subgroups

$$H^{\overline{\alpha}} = \cup_{\gamma \in \overline{\alpha}} G^\gamma \text{ and } H_{\overline{\alpha}} = \cap_{\gamma \in \overline{\alpha}} G_\gamma.$$

We make a study of these groups. For example, $H_{\overline{\alpha}} = \bigvee \mathcal{P} = \bigvee S_\gamma$ for all $\gamma \leq \alpha$, where $\mathcal{P}$ is the set of all minimal primes contained in G_γ. G is *normal valued* (i.e., $G_\gamma \triangleleft G^\gamma$ for each $\gamma \in \Gamma$) iff $H_{\overline{\gamma}} \triangleleft H^{\overline{\gamma}}$ for each $\overline{\gamma} \in \overline{\Gamma}$. G is finite valued iff $H^\gamma = S\gamma \boxplus S_\gamma'$ for each $\gamma \in \Gamma$.

In section 5 all of the above is applied to abelian ℓ-groups, and we obtain some very interesting representations of these groups.

Notation and Definitions. If $\{A_\lambda | \lambda \in \Lambda\}$ is a set of ℓ-groups then ΠA_λ (ΣA_λ) will denote the cardinal product (sum) of the groups A_λ. $G(a) = \{g \in G|\ |g| \leq n|a|$ for some $n > 0\}$ is the convex ℓ-subgroup of G generated by a. An element $\gamma \in \Gamma$ is *essential* if there exists $0 \neq g \in G$ with all its values $\leq \gamma$. A lattice ordered ring is an *f-ring* if $a \wedge b = 0$ and $c > 0$ imply $ac \wedge b = ca \wedge b = 0$.

2. THE CONVEX ℓ-SUBGROUPS S_γ OF G.

Throughout let G be an ℓ-group and let Γ be the root system of all the regular subgroups of G. If Λ is an ideal in Γ then

$$\begin{aligned}\Lambda^* &= \{g \in G \mid \text{each value of } g \text{ belongs to } \Lambda\}\\ &= \cap \{G_\gamma | \gamma \in \Gamma \setminus \Lambda\}\end{aligned}$$

In particular, $\Lambda^* \in \mathcal{C}(G)$. Now for each $\gamma \in \Gamma$ let

$$S_\gamma = \{g \in G | \text{no value of } g \text{ is comparable with } \gamma\}$$

Since the set of all the elements in Γ not comparable with γ is an ideal we have $S_\gamma \in \mathcal{C}(G)$.

LEMMA 2.1. If $\alpha, \delta_1, \delta_2, \ldots, \delta_n$ is a trivially ordered subset of Γ then there exists $0 < c \in (G^\alpha \setminus G_\alpha) \cap (\cap_{i=1}^n S_{\delta_i})$.

Proof. If $0 < x \in G^\alpha \setminus G_\alpha$ and $0 < y \in G_{\delta_1} \setminus G_\alpha$ then $0 < x \wedge y \in (G^\alpha \setminus G_\alpha) \cap G_{\delta_1}$. Now pick $0 < a \in (G^\alpha \setminus G_\alpha) \cap G_{\delta_1}$ and

$0 < b \in (G^{\delta_1} \setminus G_{\delta_1}) \cap G_\alpha$ and consider

$$c = (a - b) \vee 0 = a - (a \wedge b)$$

Since $a \wedge b \in G_\alpha$, $c \in (G^\alpha \setminus G_\alpha) \cap G_{\delta_1}$. If $\delta_1 > \gamma$ a value of c then $G_\gamma < G_\gamma + c = G_\gamma + (a - b) \vee 0 = G_\gamma + a - b \vee G_\gamma$ so $G_\gamma + a - b > G_\gamma$ and hence $G_{\delta_1} = G_{\delta_1} + a \geq G_{\delta_1} + b > G_{\delta_1}$, a contradiction. Therefore $c \in (G^\alpha \setminus G_\alpha) \cap S_{\delta_1}$. So for each i pick $0 < c_i \in (G^\alpha \setminus G_\alpha) \cap S_{\delta_i}$ then

$$c_1 \wedge c_2 \wedge \ldots \wedge c_n \in (G^\alpha \setminus G_\alpha) \cap (\cap_{i=1}^n S_{\delta_i}).$$

THEOREM 2.2. $S_\gamma = \cap$ all the minimal prime subgroups contained in G_γ
$= \cup\{a' \mid 0 < a \in G \setminus G_\gamma\}$
$= \cap\{G_\delta \mid \delta$ and γ are comparable$\}$.
$S'_\gamma = \cap\{a'' \mid 0 < a \in G \setminus G_\gamma\}$.
$G_\delta \supseteq S_\gamma$ if and only if δ and γ are comparable, and $\alpha \| \beta$ if and only if $S_\alpha \| S_\beta$.

Proof. In [11] it is shown that $T = \cup\{a' \mid 0 < a \in G \setminus G_\gamma\} = \cap$ all the minimal primes contained in G_γ. If $0 < x \in T$ then $x \wedge a = 0$ for some $0 < a \in G \setminus G_\gamma$ so $x \in G_\gamma$. If α is a value of x then clearly $\alpha \not> \gamma$ and if $\alpha < \gamma$ then x, a $\notin G_\alpha$ but $x \wedge a = 0$, which is impossible. Therefore $T \subseteq S_\gamma$.

Now consider $0 < x \in S_\gamma$ and suppose (by way of contradiction) that $x \wedge a > 0$ for each $0 < a \in G \setminus G_\gamma$. Then the set of all the finite intersections of elements in $\{x\} \cup \{a \mid 0 < a \in G \setminus G_\gamma\}$ does not contain 0 so it is contained in an ultra filter K of G^+. But $\overline{K} = \cup\{k' \mid k \in K\}$ is a minimal prime (see [11]) contained in G_γ that does not contain x. Thus a value of x contains $\overline{K}$ and hence must be comparable with γ, a contradiction. It follows that $S_\gamma \subseteq T$.

If δ and γ are comparable then $G_\delta \supseteq P$ a minimal prime contained in G_γ so $G_\delta \supseteq P \supseteq S_\gamma$. If $\delta \| \gamma$ then there exists $0 < x \in (G^\delta \setminus G_\delta) \cap S_\gamma$ so $x \in S_\gamma \setminus G_\delta$. Thus $G_\delta \supseteq S_\gamma$ iff δ and γ are comparable and hence $S_\gamma = \cap\{G_\delta \mid \delta$ and γ are comparable$\}$ since each convex ℓ-subgroup is the intersection of regular subgroups.

Finally if $S_\alpha \supseteq S_\beta$ then $G_\alpha \supseteq S_\alpha \supseteq S_\beta$ so α and β are comparable and, if $\alpha < \beta$ then clearly $S_\alpha \supseteq S_\beta$.

COROLLARY 1. G is representable if and only if each $S_\gamma \triangleleft G$.

Proof. ($\rightarrow$) Each minimal prime is normal and hence each $S_\gamma \triangleleft G$.

($\leftarrow$) Since $G_\gamma \supseteq S_\gamma$ we have $-g + G_\gamma + g \supseteq S_\gamma$ for each $g \in G$ and so $-g + G_\gamma + g$ and G_γ are comparable. Thus (see [5]) G is representable.

A convex ℓ-subgroup C of G is a *z-subgroup* if $0 < x \in C$ implies that $x'' \subseteq C$. Each minimal prime is a z-subgroup, and if G is an f-ring then each z-subgroup is a ring ideal.

COROLLARY 2. Each S_γ is a z-subgroup so if G is an f-ring then each S_γ is a ring ideal.

COROLLARY 3. G is hyperarchimedian if and only if $G_\gamma = S_\gamma$ for each $\gamma \in \Gamma$.

Proof. G_γ is a minimal prime iff $G_\gamma = S_\gamma$.

THEOREM 2.3. For a convex ℓ-subgroup C of G and $0 < g \in G \setminus C$ the following are equivalent.

1) There is a unique value of g that contains C and $0 < x \notin C$ implies $x \wedge g \notin C$.
2) $G_\gamma \supseteq C \supseteq S_\gamma$ for a value γ of g.

If $C \triangleleft G$ then each of the above is equivalent to

3) $C + g$ is not in the lex-kernel of G/C.

Proof. ($1 \rightarrow 2$) Let G_γ be the unique value of g that contains C and suppose (by way of contradiction) that $C \subseteq G_\alpha$ for some $\alpha || \gamma$. Pick $0 < x \in (G^\alpha \setminus G_\alpha) \cap S_\gamma$. Then $g \wedge x \notin C$ so $C \subseteq G_\delta$ a value of $g \wedge x$. Now $g \notin G_\delta$ so $\delta \leq \gamma$ and since $x \notin G_\delta$ it must have a value $\geq \delta$ and hence comparable with γ, but this contradicts the fact that $x \in S_\gamma$. Thus

$$C = \bigcap \text{ some of the } C_\alpha \text{ with } \alpha \text{ comparable with } \gamma$$
$$\supseteq \bigcap\{G_\delta \mid \delta \text{ and } \gamma \text{ are comparable}\} = S_\gamma.$$

($2 \rightarrow 1$) If $G_\beta \supseteq C \supseteq S_\gamma$ then β and γ are comparable so G_γ is the unique value of g that contains C. Now $C = \bigcap G_\delta$ where each $G_\delta \subseteq G_\gamma$ so if $0 < x \notin C$ then $x \notin G_\delta$ for some δ and hence $x \wedge g \notin C_\delta \supseteq C$.

If A is an ℓ-group with lex-kernel $K \subset A$ then $0 < a \in A/K$ iff a is special and a unit.

$(1 \rightarrow 3)$ G^{γ}/C is the only value of $C + g$ in G/C and clearly $C + g$ is a unit.

$(3 \rightarrow 1)$ $C + g$ is special in G/C so g has a unique value that contains C and since $C + g$ is a unit in G/C we have (1).

We shall say that an ℓ-group G is a *lex-group* if it is not equal to its lex kernel or equivalently if G is a proper lex-extension of an ℓ-ideal.

COROLLARY 1. For an ℓ-ideal C of G the following are equivalent.

a) G/C is a lex group.

b) $G_\gamma \supseteq C \supseteq S_\gamma$ for some $\gamma \in \Gamma$.

Thus G is a lex-group if and only if $S_\gamma = 0$ for some $\gamma \in \Gamma$. If G is representable then each G/S_γ is a lex-group.

COROLLARY 2. If G is representable and $0 \neq g \in G$ then C is a minimal ℓ-ideal such that $C + g$ is not in the lex kernel of G/C if and only if $C = S_\gamma$ for some value γ of g.

COROLLARY 3. If G is abelian then G/S_γ is a lex-group with kernel G_γ/S_γ. Moreover S_γ is the smallest ℓ-ideal such that G/S_γ is a lex-extension of G_γ/S_γ.

PROPOSITION 2.4. If $A \in \mathscr{C}(G)$ then

1) $\{\alpha \in \Gamma \mid G_\alpha \not\supseteq A\} = \{\alpha \in \Gamma \mid \alpha$ is the value of some $a \in A\}$.

2) $A' = \cap \{S_\alpha \mid G_\alpha \not\supseteq A\}$.

Proof. (1) $G_\alpha \not\supseteq A$ iff $G_\alpha \cap A$ is regular in A iff $A \cap (G^\alpha \setminus G_\alpha) \neq \emptyset$.

(2) If $0 < x \in A'$ and $0 < a \in A$ with value α then since $x \wedge a = 0$ no value of x is comparable with α. Thus $x \in S_\alpha$ and $A' \subseteq \cap S_\alpha$. Conversely if $0 < x \in \cap S_\alpha$ and $0 < a \in A$ then no value of x is comparable with a value of a so $x \wedge a = 0$. Thus $x \in A'$ and $\cap S_\alpha \subseteq A'$.

COROLLARY. $S_\gamma' = \cap_{\alpha || \gamma} S_\alpha = \cap \{G_\delta \mid \delta$ is comparable with some $\alpha || \gamma\} \subseteq \{g \in G \mid$ each value of g is comparable with $\gamma\}$.

Proof. $S_\gamma' = \cap \{S_\alpha \mid G_\alpha \not\supseteq S_\gamma\}$ and by Proposition 2.2 $G_\alpha \not\supseteq S_\gamma$ iff $\alpha || \gamma$. If α is a value of $g \in S_\gamma'$ and $\alpha || \gamma$ then $g \in S_\alpha \subseteq G_\alpha$, a contradiction.

PROPOSITION 2.5. $G^\gamma \cap S'_\gamma = G^\gamma \cap (\cap_{\delta||\gamma} G_\delta) = \{g \in G | \text{each value of } g \text{ is } \leq \gamma\}$ $= \cap G(b)$ for all $0 < b \in G^\gamma \backslash G_\gamma$. γ is essential if and only if $S'_\gamma \neq 0$.

Proof. $G^\gamma \cap S'_\gamma = G^\gamma \cap (\cap\{G_\delta | \delta \text{ is comparable with some } \alpha||\gamma \text{ and } \delta \not\geq \gamma\}) =$ $G^\gamma \cap (\cap_{\delta||\gamma} G_\delta) = \{g \in G | \text{each value of } g \text{ is } \leq \gamma\}$.

Now consider $0 < g \in G^\gamma \cap S'_\gamma$ and $0 < b \in G^\gamma \backslash G_\gamma$. If $g \in G$ then each value of g is $< \gamma$ so $g < b$. If $g \in G^\gamma \backslash G_\gamma$ then γ is special so $G_\gamma \lhd G^\gamma$ and hence $G_\gamma + g < G_\gamma + nb$ for some $n > 0$, so $g < nb$. Thus $g \in G(b)$ and $G^\gamma \cap S'_\gamma \subseteq \cap G(b)$.

Conversely consider $0 < x \in G \cap G(b) \subseteq G^\gamma$ with value α. If $\alpha \not\leq \gamma$ then $\alpha||\gamma$. Now pick $0 < b \in (G^\gamma \backslash G_\gamma) \cap G_\alpha$. Then $x \notin G_\alpha \supseteq G(b)$ a contradiction. Thus each value of x is $\leq \gamma$ so $x \in G^\gamma \cap S_\gamma'$.

If $0 < b \in G$ is special with value γ then $G^\gamma = b' \boxplus G(b)$ and $G_\gamma = b' \boxplus K$, where $K = G(b) \cap G_\gamma$ is the value of b in $G(b)$ [8].

LEMMA 2.6. If $0 < b \in G$ is special with value γ then

$$S_\gamma = b' \text{ and } S'_\gamma = b''.$$

Proof. By Proposition 2.4 $b' = G(b)' = \cap\{S_\alpha | \alpha \text{ is a value of some}$ $a \in G(b)\}$ but γ is the largest element in $\{\alpha | \alpha \text{ is a value of some } a \in G(b)\}$ so $b' = S_\gamma$.

COROLLARY 1. γ is special if and only if $G^\gamma = S_\gamma \boxplus T$.

Proof. ($\leftarrow$) If $0 < g \in G^\gamma \backslash G_\gamma$ then $g = x + y$ where $x \in S_\gamma \subseteq G_\gamma$ and $y \in (G^\gamma \backslash G_\gamma) \cap S'_\gamma$ so by Proposition 2.5 γ is the only value of y.

COROLLARY 2. G is finite valued if and only if $G^\gamma = S_\gamma \boxplus T_\gamma$ for each $\gamma \in \Gamma$.

Proof. G is finite valued if each γ is special [8].

3. THE TREE $\mathcal{S}$ OF ALL THE S_γ's.

Let $\mathcal{S} = \{S_\gamma | \gamma \in \Gamma\}$. If $\alpha < \beta$ then $S_\alpha \supseteq S_\beta$ and $\alpha||\beta$ iff $S_\alpha||S_\beta$ so $S_\alpha \supset S_\beta$ implies $\alpha < \beta$. Thus $\mathcal{S}$ is a tree with respect to the po $\subseteq$. In particular, each element in $\mathcal{S}$ exceeds at most one atom (i.e., minimal element in $\mathcal{S}$) and 0 is an atom iff G is a lex group.

LEMMA 3.1. $S_\gamma \subset S_\alpha$ iff $\gamma > \alpha$ and β with $\alpha || \beta$ for some $\beta \in \Gamma$. Thus S_α is an atom iff $\gamma > \alpha$ and β implies α and β are comparable. $\mathcal{S}$ is finite iff Γ contains a finite number of roots.

Proof. ($\rightarrow$) If $0 < x \in S_\alpha \setminus S_\gamma$ then x has a value β comparable with γ and since β is not comparable with α we have $\beta < \gamma$.

($\leftarrow$) $S_\gamma \subseteq S_\alpha \cap S_\beta \subseteq S_\alpha$ but if $S_\alpha \cap S_\beta = S_\alpha$ then $S_\alpha \subseteq S_\beta$ so α and β are comparable.
then $S_\alpha \subseteq S_\beta$ so α and β are comparable.

COROLLARY. $S_\alpha = S_\beta$ iff for each $\gamma \in \Gamma$, γ is comparable with α iff γ is comparable with β or equivalently α and β lie on the same roots in Γ.

If S_γ is not maximal in $\mathcal{S}$ then $S_\gamma = \cap S_\alpha$ for all $S_\alpha \supseteq S_\gamma$. For if $0 < a \in (\cap S_\alpha) \setminus S_\gamma$ then a has a value β comparable with γ. Either $S_\beta \supseteq S_\gamma$ and hence $a \in S_\delta \subseteq G_\beta$ or $S_\beta \subseteq S_\gamma \subset S_\delta$ and since β and δ are comparable $a \in S_\delta \subseteq G_\beta$.

PROPOSITION 3.2. For $\gamma \in \Gamma$ the following are equivalent:

1) S_γ is maximal in $\mathcal{S}$.
2) S_γ is a prime subgroup.
3) S_γ is a minimal prime subgroup.
4) $\{\gamma \in \Gamma | \alpha \leq \gamma\}$ is a chain.
5) γ is contained in a unique root in Γ.
6) G_γ contains a unique minimal prime.

Proof. Clearly $4 \leftrightarrow 5 \leftrightarrow 6$, $3 \rightarrow 2$ and by LEMMA 3.1, $1 \rightarrow 4$.

($4 \rightarrow 3$) $S_\gamma = \bigcap_{\alpha \leq \gamma} G_\alpha$ which is a minimal prime.

($2 \rightarrow 1$) Suppose that $S_\gamma \subset S_\alpha$ then by LEMMA 3.1 $\gamma > \alpha$ and β with $\alpha || \beta$. Pick $0 < a \in (G^\alpha \setminus G_\alpha) \cap G_\beta$ and $0 < b \in (G^\beta \setminus G_\beta) \cap G_\alpha$. Then $a = a \wedge b + \bar{a}$ and $b = a \wedge b + \bar{b}$ where $\bar{a} \in G^\alpha \setminus G_\alpha$ and $\bar{b} \in G^\beta \setminus G_\beta$. Thus $\bar{a}, \bar{b} \notin S_\gamma$ and $\bar{a} \wedge \bar{b} = 0$ so S_γ is not prime.

COROLLARY 1. $\mathcal{S}$ consists of atoms or equivalently $\mathcal{S}$ is trivially ordered iff each prime in G exceeds a unique minimal prime.

Note that in this case Γ is the join of disjoint chains and such groups form a torsion class [15].

COROLLARY 2. If S_γ is maximal in $\mathcal{S}$ then $G^\gamma \cap S'_\gamma$ is an o-group which is non-zero iff γ is essential.

Proof. By Proposition 2.5, $G^\gamma \cap S'_\gamma = \{g \in G | \text{all values of } g \text{ are } \leq \gamma\} \neq 0$ iff G_γ is essential and since S_γ is maximal $\{\alpha \in \Gamma | \alpha \leq \gamma\}$ is a chain. Thus $G^\gamma \cap S'_\gamma$ consists of basic elements and so it is an o-group.

COROLLARY 3. γ is basic iff γ is special and S_γ is maximal.

Proof. ($\rightarrow$) If $0 < a \in G^\gamma \backslash G_\gamma$ is basic then $G(a)$ is an o-group so $\{\alpha \in \Gamma | \alpha < \gamma\}$ is a chain.

($\leftarrow$) If $0 < a \in G^\gamma \backslash G_\gamma$ is special then $a \in G^\gamma \cap S'_\gamma$ is an o-group and so a is basic.

Example 6.4 shows that if $G^\gamma \cap S'_\gamma \neq 0$ is an o-group then S_γ need not be maximal and Example 6.5 shows that if γ is essential and S_γ is maximal then γ must not be special.

If $\alpha, \beta \in \Lambda$, a root in Γ, then $S_\alpha \subseteq G_\beta$ so $\cup_\Lambda S_\lambda \subseteq \cap_\Lambda G_\lambda$, a minimal prime. If $\cup S_\lambda \in \mathcal{S}$ then it is maximal and hence prime so $\cup S_\lambda = \cap G_\lambda$, but the converse is false, and $\cup S_\lambda$ need not equal $\cap G_\lambda$ (see Example 6.1).

LEMMA 3.3. For $G \neq 0$ and a root Λ of Γ the following are equivalent:

1) $\cup S_\lambda$ is a polar.

2) $\cap S_\lambda' \neq 0$

3) $\cap S_\lambda' = a''$ where a is basic.

If this is the case then $\cup S_\lambda = S_\alpha = a'$ a minimal prime, where α is the value of a.

Proof. ($1 \rightarrow 2$) Since $\cup S_\lambda = A' \subseteq \cap G_\lambda \neq G$ we have $\cap S_\lambda' = (\cup S_\lambda)' = A'' \neq 0$.

($2 \rightarrow 3$) If $0 < a \in \cap S_\lambda'$ then $a'' \subseteq \cap S_\lambda'$ and all the values of a are comparable with each λ and so must belong to Λ. Thus a is special and $\cap S_\lambda$ consists of special elements and hence is an o-group. Thus a is basic and the o-group $\cap S_\lambda'$ contains the maximal convex o-subgroup a'' so $\cap S_\lambda' = a''$.

$(3 \to 1)$ Let α be the value of a. Then by Corollary 3 of Proposition 3.2 S_α is maximal and by Lemma 2.5 $\bigcup S_\lambda = S_\alpha = a'$.

COROLLARY 1. Each minimal prime subgroup of G is a polar iff $\bigcap_\Lambda S_\lambda' \neq 0$ for each root Λ of Γ.

Proof. $(\leftarrow)$ $\bigcap G_\lambda \supseteq \bigcup S_\lambda = a'$ a minimal prime so $\bigcap G_\lambda = a'$.

$(\to)$ $\bigcup S_\lambda \subseteq \bigcap G_\lambda = a'$ where a is basic. Since $a \notin \bigcap G_\lambda$ the value α of a belongs to Λ so $\bigcup S_\lambda = S_\alpha = a'$.

An element $0 < g \in G$ is an *atom* if g covers 0. G is *compactly generated* if for each subset A of G for which $a = \bigvee A$ exists we have that a is the join of a finite subset of A. In [2] it is shown that G is compactly generated if each positive element exceeds an atom and each minimal prime is a polar.

COROLLARY 2. G is compactly generated iff $\bigcap_\Lambda S_\lambda'$ contains an atom of G for each root Λ in Γ.

LEMMA 3.4. $G_\alpha \supseteq \bigcap_{i=1}^n S_{\delta_i}$ if and only if α is comparable with some δ_i.

Proof. $(\leftarrow)$ If α is comparable with δ_j then $G_\alpha \supset S_{\delta_j} \supset \bigcap S_{\delta_i}$.

$(\to)$ Without loss of generality $\bigcap_{i=1}^n S_\delta = \bigcap_{i=1}^m S_{\delta_i}$ where $\delta_1, .. \delta_n$ is a trivially ordered subset of Γ. If $\alpha, \delta_1, .., \delta_m$ is trivially ordered then pick $0 < g \in (G^\alpha \backslash G_\alpha) \cap (\bigcap_{i=1}^m S_{\delta_i})$, see Lemma 2.1, so $G_\alpha \not\supseteq \bigcap_{i=1}^n S_{\delta_i}$.

PROPOSITION 3.5. If $\delta_1, .., \delta_n$ is a trivially ordered subset of Γ then the following are equivalent:

1) $\bigcap S_{\delta_i} = 0$
2) $\delta_1, ..., \delta_n$ is a maximal trivally ordered subset of Γ.
3) Each root in Γ contains one of the δ_i.
4) Each minimal prime contains one of the S_{δ_i}.

Proof. $(1 \to 2)$ Each $G_\alpha \supseteq \bigcap S_{\delta_i}$ is comparable with one of the δ_i.

$(2 \to 1)$ Each α is comparable with one of the δ_i so $G_\alpha \supseteq \bigcap S_{\delta_i}$. Therefore $0 = \bigcap G_\alpha \supseteq \bigcap S_{\delta_i}$.

$(2 \to 3)$ Consider $\lambda \in \Lambda$ a root in Γ. λ is comparable to some δ_i and if

$\delta_i \notin \Lambda$ then $\lambda > \delta_i$ and λ_1 where $\delta_i || \lambda_1$ and $\lambda_1 \in \Lambda$. Now λ_1 is comparable to some $\delta_j \neq \delta_i$ and if $\delta_j \in \Lambda$ proceed. Since there is only a finite number of δ's one of these must belong to Λ.

(3 → 4) If P is a minimal prime then $P = \cap G_\lambda$ all $\lambda \in \Lambda$ a root in Γ. Now there exists $\delta_i \in \Lambda$ so $P \subseteq G_{\delta_i}$ and hence $P \supseteq S_{\delta_i}$ since S_{δ_i} is the intersection of all the minimal primes in G_{δ_i}.

(4 → 1) $0 = \cap$ all minimal primes $\supseteq \cap S_{\delta_i}$.

COROLLARY. There exists a finite set $S_{\alpha_1}, \ldots, S_{\alpha_n}$ of maximal elements of $\mathcal{S}$ with $\cap S_{\alpha_i} = 0$ if and only if $\mathcal{S}$ is finite.

Proof. (→) Each minimal prime contains one of the S_{α_i} which is a minimal prime. Thus $S_{\alpha_1}, \ldots, S_{\alpha_n}$ are the minimal primes so Γ has only a finite number of roots and hence $\mathcal{S}$ is finite.

(←) Γ has only a finite number of roots and hence only a finite number of minimal primes each of which is a maximal element in $\mathcal{S}$.

THEOREM 3.6. For an ℓ-group G the following are equivalent:

1) There exists a maximal trivially ordered subset $\alpha_1, \ldots, \alpha_n$ of Γ.

2) There exists a finite subset of $\mathcal{S}$ with zero intersection.

3) There exists a finite set of special elements, which is also a maximal disjoint subset of G.

4) Each pairwise disjoint set of roots in Γ is finite.

If this is the case then $\cap_{i=1}^{n} S_{\alpha_i} = 0$ and $0 \neq (G^{\alpha_j} \setminus G_{\alpha_j}) \cap (\cap_{i \neq j} S_{\alpha_i})$ is the set of all special elements with value α_j. If $0 < a_i$ is special with value α_i then $a_1, \ldots, a_n$ is a maximal disjoint subset of G. Moreover a_i'' is an o-group if and only if S_{α_i} is maximal in $\mathcal{S}$.

If $b_1, \ldots, b_t$ is a set of special elements of G with values $\beta_1, \ldots, \beta_t$ which is also a maximal disjoint subset of G, then $\beta_1, \ldots, \beta_t$ is a maximal trivially ordered subset of Γ.

Proof. 1 ↔ 2 ↔ 4 and (1) implies $\cap_{i=1}^{n} S_{\alpha_i} = 0$ by Proposition 3.5.

(1 → 3) Pick $0 < a \in (G^{\alpha_1} \setminus G_{\alpha_1}) \cap (\cap_{i=2}^{n} S_{\alpha_i})$. If a has a value $\beta \neq \alpha_1$ then $\beta || \alpha_1$ so β is comparable to one of the α_i and hence $a \notin S_{\alpha_i}$, a contradiction. Thus a is special with value α_1. Now if $0 < a_i$ is special with value α_i for i=1,...,n and $a \wedge a_i = 0$ for all i then for each value α of a,

$\alpha||\alpha_i$ for all i so a must be zero and hence $a_1,..,a_n$ is a maximal disjoint subset of G. By Corollary 3 of Proposition 3.2 S_{α_i} is maximal iff a_i is basic iff a_i'' is an o-group.

(3 → 1) Let $b_1,..,b_t$ be as described and suppose that $\alpha, \beta_1,.., \beta_t$ is trivially ordered. Pick $0 < a \in (C^\alpha \backslash C_\alpha) \cap (\cap_{i=1}^t S_{\beta_i})$, then a, $b_1,..,b_t$ are disjoint.

Let $\alpha_1,...,\alpha_n$ be a maximal trivially ordered subset of Γ and for each i pick $0 < a_i \in G^{\alpha_i} \backslash G_{\alpha_i}$ special. Then $a_1,...,a_n$ is a maximal disjoint subset of G and in [10] it is proven that G is a *lex-sum* of the groups $a_1'',...,a_n''$. This means that there exists a chain $A(0) \subseteq A(1) \subseteq ... \subseteq A(s)$ of convex ℓ-subgroups of G such that $G = \cup A(i)$ and $A(i) = a(i)_1'' \boxplus ... \boxplus a(i)_{t_i}''$, where each $a(i)_j$ is a special element in G. Moreover:

a) $A(0) = a_1'' \boxplus ... \boxplus a_n''$ (that is $t_o = n$ and $a(0)_i = a_i$), and

b) $a(i+1)_j'' = a(i)_k''$ for some k or $a(i+1)_j''$ is a proper lex-extension of two or more of the components of A(i).

If in addition $A(0) \triangleleft G$ then G is a *normal lex-extension* of $a_1'',...,a_n''$.

Note that if G is representable then each polar is normal in G so $A(0) \triangleleft G$. Also if A(0) is normal in G then each inner automorphism of G induces a permutation on the irreducible components $a_1'',..,a_n''$ of A(0). It follows that it must also permute the $a(1)_i''$ so $A(1) \triangleleft G$ and hence each $A(i) \triangleleft G$.

THEOREM 3.7. G is a lex sum of ℓ-groups $a_1'',...,a_n''$ where each a_i is special if and only if there is a finite subset of $\mathcal{S}$ with zero intersection.

In general $a_1'' \boxplus ... \boxplus a_n''$ need not be normal, but Byrd [4] has shown that the set S of all conjugates of the a_i'' is finite. Thus G is a normal lex sum of the minimal element in S.

It follows quite easily that $G = a_1'' \boxplus ... \boxplus a_n''$ where the a_i are special iff the corresponding S_{α_i} are atoms in $\mathcal{S}$ so iff $\mathcal{S}$ is atomic and has only a finite number of atoms. We shall obtain this result again as a corollary of Theorem 3.10.

FINITE BASIS THEOREM: For an ℓ-group G the following are equivalent.

1) G is a lex sum of a finite number of o-groups.
2) There exists a finite set of maximal elements of $\mathcal{S}$ with zero intersection.

3) $\mathcal{S}$ is finite.

4) Γ contains only a finite number of roots.

Now let $\alpha_1,\alpha_2,\ldots,\alpha_n$ be a trivially ordered subset of Γ and for each $j = 1,\ldots,n$ pick $g_j \in (G^{\delta_j} \setminus G_{\delta_j}) \cap (\cap_{i\neq j} S_{\alpha_i})$. Let $S = \cap_{i=1}^{n} S_{\alpha_i}$.

THEOREM 3.8. If $S \triangleleft G$, and this is the case if G is representable, then each $S + g_i$ is special and G/S is a lex sum of the groups $(S + g_1)''$, $\ldots,(S + g_n)''$. Moreover $(S + g_i)''$ is an o-group if and only if S_{α_i} is maximal in $\mathcal{S}$.

Proof. By Lemma 3.4 $G_\delta \supseteq S$ iff δ is comparable with some α_i so $G_{\alpha_1}/S,\ldots,G_{\alpha_n}/S$ is a maximal trivially ordered subset of $\Gamma(G/S)$ and S_{α_j}/S is the intersection of all the regular subgroups of G/S that are comparable with G_{α_j}/S. Thus by Theorem 3.6 each $S + g_i$ is special in G/S and $S + g_1$, $\ldots,S + g_n$ is a maximal disjoint subset of G/S. Therefore G/S is a lex sum of the groups $(S + g_i)''$.

Now S_{α_i} is maximal in $\mathcal{S}$ iff S_{α_i} is a minimal prime so iff S_{α_i}/S is a minimal prime in G/S. Thus by Theorem 3.6, S_{α_i} is maximal in $\mathcal{S}$ iff $(S + g_i)''$ is an o-group.

COROLLARY. If G is representable and $S_{\alpha_1},\ldots,S_{\alpha_n}$ is a trivially ordered set of maximal elements in $\mathcal{S}$ then $G/\cap S_{\alpha_i}$ has a finite basis of n elements.

PROPOSITION 3.9. For $\alpha \in \Gamma$ the following are equivalent:

1) $G = S_\alpha' \boxplus S_\alpha$.

2) $G = a'' \boxplus a'$ where a is special with value α.

3) S_α is an atom in $\mathcal{S}$ and each $\beta \geq \alpha$ is special.

If this is the case then $S_\alpha' = a''$ and $S_\alpha = a'$. Moreover S_α' is an o-group if and only if S_α is maximal in $\mathcal{S}$. Condition (3) is an invariant of the lattice $\mathcal{C}(G)$ of all the convex ℓ-subgroups of G.

Proof. $(1 \to 3)$ If α is not an atom then $\gamma > \alpha$ and β with $\alpha||\beta$. Pick $0 < b \in (G^\beta \setminus G_\beta) \cap S_\alpha$ and $0 < g \in G^\gamma \setminus G$. Then $g = x + y \in S_\alpha' \boxplus S_\alpha$ and $x \in G_\beta$ for otherwise $x \wedge b \neq 0$. Thus $y \in G^\gamma \setminus G_\gamma$ so $y \notin S_\alpha$ a contradiction. Therefore S_α is an atom.

If $0 < g \in G^\delta \setminus G_\delta$ with $\delta \geq \alpha$ then $g = x + y \in S'_\alpha \boxplus S_\alpha$. Since $y \in S_\alpha \subseteq G_\alpha$ and $G_\alpha \subseteq G_\delta$, $x \in G^\delta \setminus G_\delta$, but since $S_\alpha = S_\delta$, $x \in G^\delta \cap S_\delta'$ and so by Proposition 2.5 all the values of x are $\leq \delta$. Thus x is special with value δ.

$(3 \rightarrow 2)$ Pick $0 < a \in G^\alpha \setminus G_\alpha$ special and suppose that $0 < b \in G \setminus (a'' \boxplus a')$. Then $b > a''$ (see [13]) so b has a value $\beta > \alpha$ and we may assume b is special with value β. Now $b > a''$ a maximal lex-subgroup so $b > a$, $c > 0$ with $a \wedge c = 0$ (see [10]). Let γ be a value of c, then $\beta > \alpha$, γ and $\alpha \| \gamma$ so $S_\beta \subset S_\alpha$ a contradiction. Therefore $G = a'' \boxplus a'$.

Now $S_\alpha' = a''$ is an o-group iff G(a) is an o-group iff the regular subgroups of G(a) form a chain iff the elements below α in Γ form a chain iff S_γ is maximal in $\mathcal{S}$ (Proposition 3.2).

Finally to show that whether or not condition (3) holds can be determined by examining $\mathcal{C}(G)$ it suffices to show.

LEMMA: A regular subgroup C of G is special if and only if:

$$\bigcap_{G_\gamma \nsubseteq C} G_\gamma = \Big(\bigcap_{G_\gamma \nsubseteq C^\gamma} G_\gamma\Big) \cap C \subset \bigcap_{G_\gamma \nsubseteq C} G_\gamma$$

Thus the special subgroups can be distinguished in the lattice $\mathcal{C}(G)$.

Proof. If C is the only value of g then $g \in (\bigcap_{G_\gamma \nsubseteq C} G_\gamma) \setminus C$. Conversely suppose $g \in (\bigcap_{G_\gamma \nsubseteq C} G_\gamma) \setminus (\bigcap_{G_\gamma \nsubseteq C} G_\gamma)$ then $g \notin C$ and since $g \in \bigcap_{G_\gamma \nsubseteq C} G_\gamma$ all values of g are contained in C so C is the unique value of g.

COROLLARY 1. If G is finite valued then $G = S_\alpha' \boxplus S_\alpha$ if and only if S_α is an atom in $\mathcal{S}$.

Proof. Each $\gamma \in \Gamma$ is special.

Now by Corollary 3 of Proposition 3.2 γ is basic iff γ is special and S_γ is maximal.

COROLLARY 2. The regular subgroups of G that are basic can be distinguished in $\mathcal{C}(G)$.

The local structure theorem [8] fits in here. If $G = G(g)$ where g is finite valued, then $g = g_1 + \cdots + g_n$ where the g_i are disjoint and special with values $\gamma_1, \ldots, \gamma_n$.

Then:

$$G = G(g_1) \boxplus \ldots \boxplus G(g_n) = S'_{\gamma_1} \boxplus \ldots \boxplus S'_{\gamma_n} \simeq G/S_{\gamma_1} \boxplus \ldots \boxplus G/S_{\gamma_n}.$$

For $G = G(g_1) + G(g_2 + \ldots + g_n) = g_1'' \boxplus g_1' = S_{\gamma_1}' \boxplus S_{\gamma_1}$.

Let $\{S_\lambda | \lambda \in \Lambda\}$ be the set of all the (distinct) elements in $\mathcal{S}$ that are cardinal summands of G. Then for each λ

$$G = S_\lambda' \boxplus S_\lambda = a_\lambda'' \boxplus a_\lambda'$$

where a_λ is special with value λ. Thus for $g \in G$ we have $g = g_\lambda + g^\lambda \in a_\lambda'' \boxplus a_\lambda'$ so the mapping

$$g \xrightarrow{\sigma} (\ldots, g_\lambda, \ldots)$$

is an ℓ-homomorphism of G into $\Pi a_\lambda''$ with kernel $\cap S_\lambda$.
Moreover

$$\Sigma\, a_\lambda'' \subseteq G\sigma \subset \Pi a_\lambda''.$$

THEOREM 3.10. The set of all elements in $\mathcal{S}$ that are cardinal summands of G has zero intersection if and only if there is an ℓ-isomorphism σ of G such that

$$\Sigma\, a_\lambda'' \subseteq G\sigma \subseteq a_\lambda''$$

where the a_λ are special elements of G. If this is the case then the a_λ'' are all the indecomposable summands of G and the a_λ' are all the summands of G in $\mathcal{S}$. In particular, $G = \Sigma a_\lambda''$ if and only if $\mathcal{S}$ is atomic and each atom is a summand. In this case the atoms of $\mathcal{S}$ are the a_λ'. a_λ'' is an o-group if and only if a_λ' is maximal in $\mathcal{S}$.

Proof. The first statement follows from the above and Proposition 3.9. Suppose that $G = A \boxplus B$ where $0 \neq A$ is indecomposable. Then since $G = a_\lambda'' \boxplus a_\lambda'$ we have $A = A \cap a_\lambda'' \boxplus A \cap a_\lambda'$. If $A \cap a_\lambda'' = 0$ for all λ then $A \cap \Sigma a_\lambda'' = 0$, a contradiction. Thus $A \cap a_\lambda'' \neq 0$ for some λ so $A = A \cap a_\lambda''$ and hence $A \subseteq a_\lambda''$. If $A \subset a_\lambda''$ then $a_\lambda'' = A \boxplus (\alpha_\lambda'' \cap B)$ which contradicts the fact that a_λ is special. Thus the a_λ'' are the indecomposable summands of G. If $G = S_\gamma' \boxplus S_\gamma$ then by Proposition 3.9 S_γ' is indecomposable so it must be one of the a_λ'' and hence $S_\gamma = a_\lambda'$.

If $\mathcal{S}$ is atomic and each atom is a summand then we may assume that $\Sigma a_\lambda'' \subseteq G \subseteq \Pi a_\lambda''$. If $G \supset \Sigma a_\lambda''$ then $\Sigma a_\lambda'' \subseteq G_\gamma$ for some $\gamma \in \Gamma$ and we may assume that S_γ is an atom and hence special. Pick $0 < a \in G^\gamma \backslash G_\gamma$ special. Then a'' is one of the a_λ'' so $a \in a'' = a_\lambda'' \subseteq G_\gamma$, a contradiction. Finally if $G = \Sigma a_\lambda''$ then Γ is of the form

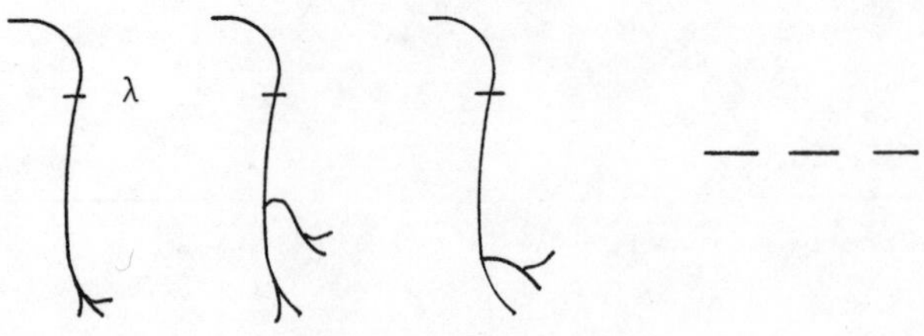

where the elements above each λ are special, so clearly $\mathcal{S}$ is atomic and each atom a_λ' is a summand.

COROLLARY 1. If G is finite valued then $\mathcal{S}$ is atomic if and only if $G = \Sigma a_\lambda''$ where each a_λ is special.

Proof. This follows from the fact that S_γ is a summand iff S_γ is an atom (Corollary 1 of Proposition 3.9).

COROLLARY 2. G is a cardinal sum of o-groups if and only if each S_γ is a summand.

Proof. (→) If $G = \Sigma T_\lambda$, where each T_λ is an o-group then clearly each S_γ is of the form ΣT_λ all $\lambda \neq \gamma$.

(←) Each S_γ is a summand and hence an atom so $G = \Sigma a_\lambda''$. But since all the S_γ are atoms they are also maximal in $\mathcal{S}$ so each a_λ'' is an o-group.

Note that this is a torsion class of ℓ-groups.

COROLLARY 3. $G = a_1'' \boxplus \dots \boxplus a_n''$ where the a_i are special if and only if $\mathcal{S}$ is atomic and contains only a finite number of atoms.

Proof. (→) $a_1', \dots, a_n'$ are the atoms in $\mathcal{S}$.

(←) Let $S_{\alpha_1}, S_{\alpha_2}, \dots, S_{\alpha_n}$ be the atoms and consider $\beta \geq \alpha_1$. Then $\beta, \alpha_2, \dots, \alpha_n$ is a maximal trivially ordered subset of Γ so β is special (see Theorem 3.6) and hence by Proposition 3.9 S_{α_1} is a summand. Similarly each

S_{α_i} is a summand so by the Theorem $G = a_1'' \boxplus \ldots \boxplus a_n''$ where a_i is special with value α_i.

COROLLARY 4. (Finite Basis Theorem) If G has a finite basis or equivalently Γ has only a finite number of roots, then G is a lex-group or $G = A \boxplus B$ with $A \neq 0 \neq B$. Thus by induction G is a lexicographic sum of o-groups.

Proof. $\mathcal{S}$ is finite so atomic and contains only a finite number of atoms. Thus by the last Corollary $G = S_{\gamma_1}' \boxplus \ldots \boxplus S_{\gamma_n}'$ where the S_{γ_i} are the atoms in $\mathcal{S}$. If $G = S_{\gamma_1}'$ then G is a lex group.

4. THE ROOT SYSTEM Γ AND THE SUBGROUPS $H^{\overline{\gamma}}$ AND $H_{\overline{\gamma}}$.

For $\alpha, \beta \in \Gamma$ we define $\alpha \sim \beta$ if $S_\alpha = S_\beta$. Then $\sim$ is an equivalence relation and we shall denote the equivalence class that contains α by $\overline{\alpha}$ and the set of all equivalence classes by $\overline{\Gamma}$. Note that $\overline{\Gamma}$ contains only a finite number of roots iff G has a finite basis. Also we have shown $\alpha \sim \beta$ iff $\{G_\delta | \delta$ and α are comparable$\} = \{G_\delta | \delta$ and β are comparable$\}$ iff G_α and G_β contain the same set of minimal primes iff α and β lie on the same roots in Γ. In particular, each $\overline{\alpha}$ is a convex chain in Γ with at most one branch point at the lower end.

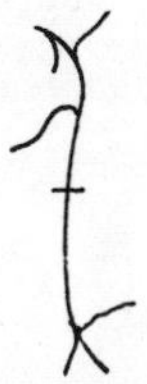

α

Define $\overline{\beta} > \overline{\alpha}$ if $\overline{\beta} \neq \overline{\alpha}$ and $\beta > \alpha$ or equivalently if $\beta > \alpha$ and γ with $\alpha || \gamma$. Then $\overline{\Gamma}$ is a root system where $\overline{\beta} > \overline{\alpha}$ implies $\overline{\beta} > \overline{\gamma}$ with $\overline{\alpha} || \overline{\gamma}$ and the map $\alpha \to \overline{\alpha}$ is an o-homomorphism of Γ onto $\overline{\Gamma}$. Also $\alpha || \beta$ iff $\overline{\alpha} || \overline{\beta}$.

Now each o-permutation $\alpha \to \alpha^*$ of Γ preserves this equivalence relation so $\overline{\alpha^*} = \overline{\alpha}^* = \{\beta \in \Gamma | \beta = \delta^*$ for some $\delta \in \overline{\alpha}\}$ and $\overline{\alpha} \to \overline{\alpha^*}$ is an o-permutation of $\overline{\Gamma}$.

If τ is an o-homomorphism of Γ onto a po set Δ such that $\gamma || \beta$ iff $\alpha\tau || \beta\tau$ then Δ is a root system and for each $\delta \in \Delta$, $\{\gamma \in \Gamma | \gamma\tau = \delta\} \subseteq \overline{\alpha}$ for some $\alpha \in \Gamma$. Thus there exists a unique o-homomorphism π of Δ onto $\overline{\Gamma}$ such that $\gamma\tau\pi = \overline{\alpha}$. We make no use of this result and so shall omit the straightforward proof.

For each $\overline{\alpha} \in \overline{\Gamma}$ define

$$H^{\overline{\alpha}} = \bigcup_{\gamma \in \overline{\alpha}} G^{\gamma} \text{ and } H_{\overline{\alpha}} = \bigcap_{\gamma \in \overline{\alpha}} G_{\gamma}$$

Since $\overline{\alpha}$ is a chain $H_{\overline{\alpha}}$ is prime, also it is easy to show that

a) $\overline{\alpha} < \overline{\beta}$ iff $H^{\overline{\alpha}} \subseteq H_{\overline{\beta}}$.
b) If $0 \neq g \in G$ then $g \in H^{\overline{\alpha}} \setminus H_{\overline{\alpha}}$ for some $\overline{\alpha}$.
c) If $g \notin H^{\overline{\alpha}}$ then $g \in H^{\overline{\beta}} \setminus H_{\overline{\beta}}$ for some $\overline{\beta} > \overline{\alpha}$ so $H^{\overline{\alpha}} = \bigcap \{H_{\overline{\beta}} \mid \overline{\beta} > \overline{\alpha}\}$.

PROPOSITION 4.1. For each $\gamma \in \Gamma$, $H_{\overline{\gamma}} = \bigvee \mathcal{P} = \bigvee S_{\alpha}$ for all $\alpha \leq \gamma$, where $\mathcal{P}$ is the set of all minimal primes in G_{γ}.

Proof. If $\delta \in \overline{\gamma}$ then G_{δ} and G_{γ} exceed the same set of minimal primes so $H_{\overline{\gamma}} = \bigcap G_{\delta} \supseteq \bigvee \mathcal{P}$. If $\alpha \leq \delta$ then $S_{\alpha} = \bigcap$ minimal primes in $G_{\alpha} \subseteq$ some $P \in \mathcal{P}$ so $\bigvee \mathcal{P} \supseteq S_{\alpha}$. Next $\bigvee S_{\alpha}$ is prime. For if $0 = a \wedge b$ and neither belongs to S_{δ} then a(b) has a value $\mu(\nu)$ where $\gamma > \mu$ and ν and $\mu || \nu$. But then $a \in S_{\nu}$, and $S_{\nu} \subseteq \bigvee S_{\alpha}$. Thus it follows that $a \in \bigvee S_{\alpha}$ or $b \in \bigvee S_{\alpha}$.

If $H_{\overline{\gamma}} \supset \bigvee S_{\alpha}$ then since they are both prime we have $H_{\overline{\gamma}} \supset G_{\mu} \supseteq \bigvee S_{\alpha}$ and so $\gamma > \mu$ and ν with $\mu || \nu$. Pick $0 < x \in (G^{\mu} \setminus G_{\mu}) \cap S_{\nu}$. Then $x \in \bigvee S_{\alpha} \subseteq G_{\mu}$ a contradiction.

COROLLARY 1. If G is an f-ring then each $H_{\overline{\gamma}}$ is a ring ideal and so is each $H^{\overline{\gamma}}$.

COROLLARY 2. For an ℓ-group G the following are equivalent.

1) G is representable.
2) $H_{\overline{\gamma}} \triangleleft G$ for each $\overline{\gamma} \in \overline{\Gamma}$,
3) Each inner automorphism of G induces the identity permutation on $\overline{\Gamma}$.

Proof. This follows from the fact that G is representable iff each $S_{\gamma} \triangleleft G$ (Corollary 1 of Proposition 2.2).

PROPOSITION 4.2. G is normal valued if and only if $H_{\overline{\gamma}} \triangleleft H^{\overline{\gamma}}$ for each $\overline{\gamma} \in \overline{\Gamma}$.

Proof. If $H_{\overline{\gamma}} \lhd H^{\overline{\gamma}}$ then G^{γ} covers G_{γ} in the o-group $H^{\overline{\gamma}} \setminus H_{\overline{\gamma}}$ so $G_{\gamma} \lhd G^{\gamma}$. If G is normal valued and $0 \neq x \in H^{\overline{\gamma}} \setminus H_{\overline{\gamma}}$ then (without loss of generality) $x \in G^{\gamma} \setminus G_{\gamma}$ and hence if $a \to a^*$ is the o-permutation of Γ induced by the inner automorphism $g \to -x + g + x$ of G then $\gamma^* = \gamma$ so $\overline{\gamma} = \overline{\gamma^*} = \overline{\gamma}^*$. Therefore $-x + H_{\overline{\gamma}} + x = H_{\overline{\gamma}^*} = H_{\overline{\gamma}}$.

Recall that $\mathcal{S}$ consists of atoms iff Γ consists of disjoint chains so iff $\overline{\Gamma}$ is trivially ordered iff each $H^{\overline{\gamma}} = G$ iff each $H_{\overline{\gamma}}$ is a minimal prime. If this is the case then it follows from the above that G is representable iff it is normal valued. Whether or not such a G is always normal valued is an open question.

Suppose that G is a representable ℓ-group and $\alpha_1,\ldots,\alpha_n$ is a trivially ordered subset of Γ. Let $H = H_{\overline{\alpha}_1} \cap \cdots \cap H_{\overline{\alpha}_n}$. Then it follows quite easily that G/H has a finite basis and that the $H_{\overline{\alpha}_i}/H$ are the minimal primes. The basic subgroup is (isomorphic to) $H^{\overline{\alpha}_1}/H_{\overline{\alpha}_1} \boxplus \cdots \boxplus H^{\overline{\alpha}_n}/H_{\overline{\alpha}_1}$ and the other o-groups used in the lex sum construction of G/H are the $H^{\overline{\beta}}/H_{\overline{\beta}}$ for those $\overline{\beta}$ that exceed some $\overline{\alpha}$. Note that:

$$\begin{aligned} S'_{\gamma} &= \cap\, \{G_{\delta} \mid \delta \text{ is comparable with some } \alpha \| \gamma\} \\ &= \cap\, \{H_{\overline{\delta}} \mid \overline{\delta} \not\leq \overline{\gamma}\} \\ S_{\delta} &= \cap\, \{H_{\overline{\delta}} \mid \overline{\delta} \leq \overline{\gamma}\} \end{aligned}$$

LEMMA 4.3. a) If $x \in H_{\overline{\gamma}}$ is special then $x'' \subseteq H_{\overline{\gamma}}$

b) $H^{\overline{\gamma}} \supseteq S_{\gamma} \boxplus S'_{\gamma}$, but $S_{\gamma} \boxplus S'_{\gamma}$ need not contain $H_{\overline{\gamma}}$.

Proof. (a) If β is the value of x then $x \notin H_{\overline{\beta}}$ so $\overline{\gamma} \not\leq \overline{\beta}$. $x'' = S'_{\beta} = \cap_{\overline{\delta} \not\leq \overline{\beta}} H_{\overline{\delta}} \subseteq H_{\overline{\gamma}}$

(b) If $0 < x \in S'_{\gamma} \setminus H^{\overline{\gamma}}$ then $x \in H^{\overline{\beta}} \setminus H_{\overline{\beta}}$ for some $\overline{\beta} > \overline{\gamma}$ but $x \in S'_{\beta} = \cap_{\overline{\delta} \not\leq \overline{\gamma}} H_{\overline{\delta}} \subseteq H_{\overline{\beta}}$. Thus $H^{\overline{\delta}} \supseteq S_{\gamma} \boxplus S'_{\gamma}$.

Now let $G = C[0,1]$ and consider $G_{\gamma} = \{g \in G \mid g(\tfrac{1}{2}) = 0\}$. Then G_{γ} is not essential so $S'_{\gamma} = 0$ and $H^{\overline{\gamma}} = G \supset S_{\gamma} \boxplus S'_{\gamma} = S_{\gamma} \not\supseteq G_{\gamma} = H_{\overline{\gamma}}$.

Let $\mathcal{K}$ be the set of all the convex ℓ-subgroups of G that are joins of minimal primes and consider $K \in \mathcal{K}$.

1) $K = \cap\, \{H_{\overline{\gamma}} \mid H_{\overline{\gamma}} \supseteq K\}$

Proof. If $0 < g \in G\backslash K$ then $K \subseteq G_\gamma$ a value of g. Now all the minimal primes contained in K are also in G_γ so $K \subseteq H_{\overline{\gamma}}$.

2) If $K \supset H_{\overline{\alpha}}$ then $K \supseteq H^{\overline{\alpha}}$.

Proof. $K = \cap \{H_{\overline{\gamma}} | H_{\overline{\gamma}} \supseteq K\} \supseteq \cap \{H_{\overline{\beta}} | H_{\overline{\beta}} \supset H_{\overline{\alpha}}\} = H^{\overline{\alpha}}$.

Example 6.2 shows that $H^{\overline{\gamma}}$ need not belong to $\mathcal{K}$ so $\mathcal{K}$ need not be closed with respect to intersections of chains, but, of course, $\mathcal{K}$ is closed with respect to arbitrary joins. Example 6.5 shows that an element in $\mathcal{K}$ need not be an $H_{\overline{\gamma}}$ or an $H^{\overline{\gamma}}$. We say that K is *meet irreducible in* $\mathcal{K}$ if it is not the intersection of the elements in $\mathcal{K}$ that properly contain it.

3) The $H_{\overline{\gamma}}$'s are the meet irreducible elements in $\mathcal{K}$.

Proof. If K is meet irreducible then by (1) $K = H_{\overline{\gamma}}$, and by (2) each $H_{\overline{\gamma}}$ is meet irreducible.

Note that if K is maximal in $\mathcal{K}$ without g then each $K \subset U \in \mathcal{K}$ must contain g. Thus K is meet irreducible so $K = H_{\overline{\gamma}}$ for some value γ of g. Let

$\mathcal{N} = \{P | P$ is prime in G and $P \supset A$ a prime in G implies $P \supset B$ a prime in G with $A || B\}$.

4) $\mathcal{K} = \mathcal{N}$

Proof. If $K \in \mathcal{K}$ and $K \supset A$ a prime then there exists a minimal prime B contained in K but not in A. Thus $A||B$ so $\mathcal{K} \subseteq \mathcal{N}$. Conversely consider $P \in \mathcal{N}$ and let $\{M_i | i \in I\}$ be the set of *all* minimal primes contained in P. If $\bigvee M_i \subset P$ then $P \supset B$ a prime with $\bigvee M_i || B$. Let M be a minimal prime contained in B, then $\bigvee M_i || M$ but $M \subseteq P$ so it must be one of the M_i, a contradiction. Thus $\mathcal{N} \subseteq \mathcal{K}$.

Let $\mathcal{K}^*$ = all the intersections of subsets of $\mathcal{K}$ = all the intersections of sets of the $H_{\overline{\gamma}}$'s, and let $\mathcal{P}$ be the smallest complete sublattice of $\mathcal{C}$ that contains all the minimal primes. Then $\mathcal{K}^*$ is a complete lattice, $\mathcal{K}^* \subseteq \mathcal{P} \subseteq \mathcal{C}$, and since $\mathcal{K}^*$ contains all the minimal primes we have $\mathcal{K}^* = \mathcal{P}$ iff $\mathcal{K}^*$ is a complete sublattice of $\mathcal{C}$. In Example 6.5 $\mathcal{K} \subset \mathcal{P} = \mathcal{C}$ and $\mathcal{K}^*$ is not even a sublattice of $\mathcal{C}$. A slight variation gives $\mathcal{K}^* \subset \mathcal{P} \subset \mathcal{C}$. If G is finite valued and $\Gamma \neq \overline{\Gamma}$ then it follows from the next two theorems that $\mathcal{K}^* = \mathcal{P} \subset \mathcal{C}$. Note that the $H_{\overline{\gamma}}$'s are the meet irreducible elements in $\mathcal{K}^*$, and each S_γ and S_γ' belongs to $\mathcal{K}^*$ since both are intersections of $H_{\overline{\gamma}}$'s.

THEOREM 4.4. For an ℓ-group G the following are equivalent:

1) $\Gamma = \overline{\Gamma}$ (that is, each $\overline{\gamma} = \{\gamma\}$).
2) Each $G_\gamma \in \mathcal{K}$.
3) $\mathcal{K}^* = \mathcal{C}$.

If this is the case and G is representable (an f-ring) then each convex ℓ-subgroup of G is normal (a ring ideal).

Proof. $(1 \to 2)$ $G_\delta = H_{\overline{\delta}} \in \mathcal{K}$.

$(2 \to 3)$ This follows from the fact that each convex ℓ-group is the intersection of regular subgroups.

$(3 \to 1)$ Suppose (by way of contradiction) that $\alpha > \beta$ and $S_\alpha = S_\beta$. Each minimal prime contained in G_α is also in G_β so $G_\alpha \notin \mathcal{K}$. Now G_α is meet irreducible in $\mathcal{C}$ so it is not the intersection of elements in $\mathcal{K}$ and hence $G_\alpha \notin \mathcal{K}^*$, a contradiction.

THEOREM 4.5. If G is finite valued then $\mathcal{P} = \mathcal{K}^*$ and for $C \in \mathcal{C}$ the following are equivalent:

a) $C \in \mathcal{K}^*$
b) If $x \in C$ is special then $x'' \subseteq C$.
c) $C = \bigvee_\Lambda x_\lambda''$ where each x_λ is special.

Proof. It suffices to show that if

$$\{A_i \mid i \in I\} \subseteq \mathcal{K}^* \text{ then } \bigvee A_i \in \mathcal{K}^*.$$

If $0 < g \notin A_i$ where g is special, then $A_i \subseteq G_\gamma$ the value of g; so it suffices to show that $A_i \subseteq H_{\overline{\gamma}}$ for each $i \in I$. Now $A_i = \cap H_{\overline{\beta}}$ so $g \notin H_{\overline{\beta}}$ for some β. Thus $H_{\overline{\beta}} \subseteq G_\gamma$ and hence $A_i \subseteq H_{\overline{\beta}} \subseteq H_{\overline{\gamma}}$.

$(a \to b)$ C is the intersection of $H_{\overline{\gamma}}$'s each of which satisfies (b) by Lemma 4.3.

$(b \to c)$ $C \supseteq \bigvee x_\lambda''$ when each x_λ is special and belongs to C. If $0 < x \in C$ then $x = x_1 + \ldots + x_n$ where the x_i are disjoint special and belong to C so $x \in \bigvee x_\lambda''$.

(c → a) If γ is the value of x_λ then $x_\lambda'' = S_\lambda' \in \mathcal{K}^*$ so $\bigvee x_\lambda'' \in \mathcal{K}^*$.

The following questions are of interest:

When is $\mathcal{K}^*$ a sublattice of $\mathcal{C}$?

When is $\mathcal{K}^* = \mathcal{P}$?

When is $\mathcal{C} = \mathcal{P}$?

Describe the elements in $\mathcal{P}$.

LEMMA 4.6. Suppose that G is finite valued and let K be the lex-kernel of S_γ'.

1) $H_{\overline{\gamma}} = \bigvee S_\alpha'$ all $\overline{\alpha} \not\geq \overline{\gamma}$.
2) $K = \bigvee S_\alpha'$ all $\overline{\alpha} < \overline{\gamma}$ so $K \in \mathcal{K}^*$.
3) $S_\gamma = \bigvee S_\beta'$ all $\beta || \gamma$.

Thus $H_{\overline{\delta}} = \bigvee_{\overline{\alpha} \not\geq \overline{\gamma}} S_\alpha' = (\bigvee_{\overline{\beta} || \overline{\gamma}} S_\beta') \vee (\bigvee_{\overline{\alpha} < \overline{\gamma}} S_\alpha') = S_\gamma' \boxplus K$.

Proof. (1) Since $H_{\overline{\gamma}} \in \mathcal{K}^*$ we have $H_{\overline{\gamma}} = \bigvee x_\alpha'' = \bigvee S_\alpha'$ where each x_α is special with value γ and belongs to $H_{\overline{\gamma}}$. Now $x_\alpha \in H_{\overline{\gamma}}$ iff $\overline{\alpha} \not\geq \overline{\gamma}$.

(3) Since $S_\gamma \in \mathcal{K}^*$, $S_\gamma = \bigvee x_\beta'' = \bigvee S_\beta'$ where each x_β is special with value β and belongs to S_γ. If β is comparable with γ then $x_\beta \in G_\beta \supseteq S_\gamma$, a contradiction. If $\beta || \gamma$ then $x_\beta \wedge x_\gamma = 0$ where x_γ is special with value γ so $x_\beta \in x_\gamma' = S_\gamma$.

(2) We use the fact that $\mathcal{K}$ is the subgroup of S_γ' generated by the non-units and $S_\gamma' \backslash K$ consist of special elements that are units [8]. If $0 < x \in K$ is special with value β then $x \wedge y = 0$ for some $0 < y \in S_\gamma'$ special with value α. Now $\gamma > \alpha$ and β and $x \in y' = S_\alpha'$ so $K \subseteq \bigvee S_\alpha'$. If $\gamma > \alpha$ and β with $\alpha || \beta$, then pick $0 < a$ $(0 < b)$ special with value α (β). Then $a'' = S_\alpha' \subseteq S_\gamma'$, $b'' = S_\beta' \subseteq S_\gamma'$ and $\alpha'' \wedge b'' = 0$. Thus S_α' consists of non-units from S_γ' so $S_\gamma' \subseteq K$.

Note that it follows from (3) that $S_\gamma' = \cap_{\beta || \gamma} S_\beta$, but (see Proposition 2.4) we have shown this for an arbitrary ℓ-group. Also $(\cap S_\alpha)' \supseteq \bigvee S_\alpha'$ but equality need not hold. For if no $\gamma \in \Gamma$ is essential then each $S_\alpha' = 0$. Now take $\cap S_\alpha = 0$ then $(\cap S_\alpha)' = G \supset 0 = \bigvee S_\alpha'$.

PROPOSITION 4.7. G is finite valued if and only if $H^{\overline{\gamma}} = S_\gamma \boxplus S_\gamma'$ for each $\gamma \in \Gamma$. If this is the case then $H_{\overline{\gamma}} = S_\gamma \boxplus K$, where K is the lex-kernel of S_γ', and $S_\gamma = b'$ and $S_\gamma' = b''$ where b is special with value γ.

Moreover

$$\frac{H^{\overline{\gamma}}}{H_{\overline{\gamma}}} \simeq \frac{S_\gamma'}{K} = \frac{\bigcap S_\beta \text{ all } \beta \| \gamma}{\bigvee S_\alpha' \text{ all } \overline{\alpha} < \overline{\gamma}}$$

Proof. ($\rightarrow$) For each $\alpha \in \overline{\gamma}$ pick $0 < a_\alpha \in G^\alpha \setminus G_\alpha$ special. By Lemma 2.5 $G^\alpha = S_\gamma \boxplus G(a_\alpha)$ so $S_\gamma \boxplus \bigcup G(a_\alpha) = \bigcup G^\alpha = H^{\overline{\gamma}} \supseteq S_\gamma \boxplus S_\gamma'$ and hence $\bigcup G(a_\alpha) = S_\gamma'$.

($\leftarrow$) $G^\gamma = S_\gamma \boxplus (S_\gamma' \cap G^\gamma)$ so by Corollary to Lemma 2.5 each γ is special and hence G is finite valued.

COROLLARY. If G is finite valued abelian and divisible then each S_γ' is a direct lexicographic extension of its lex-kernel K_γ. Then $S_\gamma' = A_\gamma \oplus K_\gamma$ where A_γ is an o-subgroup of G and $H^{\overline{\gamma}}/H_{\overline{\gamma}} \simeq A$.

5. ABELIAN ℓ-GROUPS.

Let Λ be a root system and for each $\lambda \in \Lambda$ let A_λ be an o-group. Let $V(\Lambda, A_\lambda)$ be the set of all functions υ of Λ into $\bigcup A_\lambda$ where $\upsilon(\lambda) \in A_\lambda$, and the support of υ satisfies the ACC. If $\upsilon(\lambda) \neq 0$ and $\upsilon(\alpha) = 0$ for all $\lambda < \alpha \in \Lambda$ then $\upsilon(\lambda)$ is called a *maximal component* of υ. Define υ positive if each of its maximal components $\upsilon(\lambda)$ is positive in A_λ. Let

$$\Sigma(\Lambda, A_\lambda) = \{\upsilon \in V \mid \text{ support of } \upsilon \text{ is finite}\}$$
$$F(\Lambda, A_\lambda) = \{\upsilon \in V \mid \text{support of } \upsilon \text{ lies on a finite number of roots}\}$$

Then (see [7]) V is an ℓ-group with ℓ-subgroups Σ and F.

The main result in [7] asserts that if G is a divisible abelian ℓ-group then there exists an ℓ-isomorphism σ of G into $V(\Gamma, G^\gamma/G_\gamma)$ such that $g \in G^\gamma \setminus G_\gamma$ iff $(g\sigma)_\gamma$ is a maximal component of $g\sigma$, and $(g\sigma)_\gamma = G_\gamma + g$. We shall call such an isomorphism σ a *v-isomorphism*. We can choose σ so that $G \supseteq \Sigma(\Gamma, G^\gamma \setminus G_\gamma)$ provided that G is finite valued [13].

Now this can be generalized (see [9]) as follows: if G is a divisible abelian ℓ-group then there exists a v-isomorphism σ of G into $V(\overline{\Gamma}, H^\gamma/H_{\overline{\gamma}})$ such that $G\sigma \supseteq \Sigma(\overline{\Gamma}, H^\gamma/H_{\overline{\gamma}})$ provided that G is finite valued. Note that if G is finite valued then we have shown

$$\frac{H^{\overline{\gamma}}}{H_{\overline{\gamma}}} \simeq \frac{S_\gamma'}{\text{lex-kernel } S_\gamma'} \simeq A_{\overline{\gamma}}$$

where $A_{\overline{\gamma}}$ is an o-subgroup of G. Thus there is a rather natural ℓ-isomorphism of $V(\overline{\Gamma},H^{\overline{\gamma}}/H_{\overline{\gamma}})$ onto $V(\overline{\Gamma},A_{\overline{\gamma}})$.

Suppose that G is finite valued and $a < b$ special with values α,β respectively. Then $a'' \subset b''$ so $\overline{\Gamma}$ satisfies the DCC iff the lattice P of all the principal polars of G satisfies the DCC. Actually it follows from Theorem 3.11 in [7] that G is finite valued and $\overline{\Gamma}$ satisfies the DCC iff P satisfies the DCC.

For the remainder of this section assume that G is a divisible abelian ℓ-group.

THEOREM 5.1. G is finite valued and $\overline{\Gamma}$ satisfies the DCC if and only if $G \simeq \Sigma(\overline{\Gamma},H^{\overline{\gamma}}/H_{\overline{\gamma}}) = F(\overline{\Gamma},H^{\overline{\gamma}}/H_{\overline{\gamma}})$.

Proof. ($\rightarrow$) We know there exists a v-isomorphism σ such that $\Sigma \subseteq G\sigma \subseteq V$, but the support of each $g\sigma$ satisfies both the ACC and DCC and since $g\sigma$ is finite valued a straight forward argument (see [7]) shows that $g\sigma$ has finite support. Thus $G\sigma = \Sigma$ and it is easy to see that $\Sigma = F$ iff $\overline{\Gamma}$ satisfies the DCC.

($\leftarrow$) Clearly Σ is finite valued.

Now let $\underset{\sim}{F}$ be the torsion class of ℓ-groups such that each bounded disjoint subset is finite. Then $G \in \underset{\sim}{F}$ iff each bounded set in $\overline{\Gamma}$ is finite and G is finite valued.

COROLLARY. $G \in \underset{\sim}{F}$ if and only if $G \simeq \Sigma(\overline{\Gamma},H^{\overline{\gamma}}/H_{\overline{\gamma}})$ and each bounded subset of $\overline{\Gamma}$ is finite.

Note that if $\overline{\Gamma}$ =  --- and G is finite valued then $G \simeq \Sigma = F$ but $G \notin \underset{\sim}{F}$.

COROLLARY. G has a finite basis if and only if $G \simeq \Sigma(\overline{\Gamma},H^{\overline{\gamma}}/H_{\overline{\gamma}})$ with $\overline{\Gamma}$ finite (or equivalently $G \simeq \Sigma(\overline{\Gamma},H^{\gamma}/H_{\overline{\gamma}}) = V(\overline{\Gamma},H^{\gamma}/H_{\overline{\gamma}})$).

An ℓ-group is *laterally complete* if each disjoint subset has a least upper bound. Each ℓ-group A is contained in a unique minimal laterally complete ℓ-group B in which A is dense ($0 < b \in B$ implies $0 < a \leq b$ for some $a \in A$). B is the *lateral completion* of A. For a proof for representable ℓ-groups see [12] and for the general case see [1].

Suppose that the set Δ of special element of Γ is plenary (i.e., Δ is a dual ideal of Γ with zero intersection). Then there exists a v-isomorphism σ of G such that $\Sigma(\Delta,G^{\delta}/G_{\delta}) \subseteq G\sigma \subseteq V(\Delta,G^{\delta}/G_{\delta})$. This is essentially proven in [7] or see [13]. A slight modifiction gives $\Sigma(\overline{\Delta},H^{\overline{\delta}}/H_{\overline{\delta}}) \subseteq G\sigma \subseteq V(\overline{\Delta},H^{\overline{\delta}}/H_{\overline{\delta}})$. If

$\overline{\Delta}$ satisfies the DCC then (see [14]) $V(\overline{\Delta}, H^{\overline{\delta}}/H_{\overline{\delta}})$ is the lateral completion of G (and also the topological completion of G where the groups $H_{\overline{\delta}}$ form a subbasis for the topology of G).

THEOREM 5.2. If the set Δ of special elements of Γ is plenary then there exists a v-isomorphism σ of G such that

$$\Sigma(\overline{\Delta}, H^{\overline{\delta}}/H_{\overline{\delta}}) \subseteq G\sigma \subseteq V(\overline{\Delta}, H^{\overline{\delta}}/H_{\overline{\delta}}).$$

If $\overline{\Delta}$ satisfies the DCC then V is the lateral completion of G so $G \simeq V$ if and only if G is laterally complete.

Finally $\mathcal{S}$ forms a subbasis for the neighborhood of 0 of the $\mathcal{S}$-topology of G. Since $\cap\, \mathcal{S} = 0$ this topology is Hausdorff and the group and lattice operations are continuous. The following are equivalent.

1) The $\mathcal{S}$-topology is discrete.
2) G is a lex-sum of a finite number of lex-subgroups.
3) Each disjoint set of roots of Γ is finite.
4) There exists a maximal disjoint set of Γ that is finite.
5) There exists a finite set of special elements of G that is also a maximal disjoint subset of G.

One can develop a theory for the $\mathcal{S}$-topology similar to the theory of the Δ-topology given in [14] when Δ is a plenary subset of Γ. One advantage of the $\mathcal{S}$-topology is that if G is an f-ring then multiplication is also continuous.

6. EXAMPLES

EXAMPLE 6.1. Let $\Gamma =$

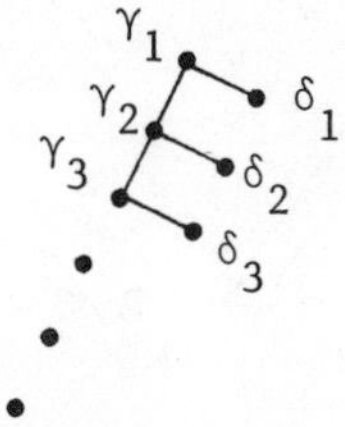

If $G = \Sigma(\Gamma, R)$ then $\cup\, S_{\gamma_i} = \cap\, G_{\gamma_i} \notin \mathcal{S}$. If $G = V(\Gamma, R)$ then $\cup\, S_{\gamma_i} = \Sigma R_{\delta_i} \subset \Pi R_{\delta_i} = \cap\, G_{\gamma_i}$. In either case $\Delta = \{\delta_i \mid i = 1, 2, \ldots\}$ is a maximal trivially ordered subset of Γ, but the minimal prime $\cap\, G_{\gamma_i}$ does not contain one of the S_{δ_i}.

EXAMPLE 6.2. Let $\Gamma =$

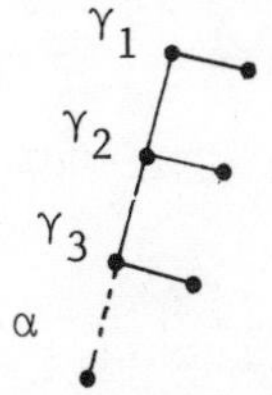

If $G = \Sigma(\Gamma,R)$ then $H^{\overline{\alpha}} = G^{\alpha} \notin \mathcal{K}$. Note also that $S_{\alpha} = \cup S_{\gamma_i}$.

EXAMPLE 6.3. Let $\Gamma =$

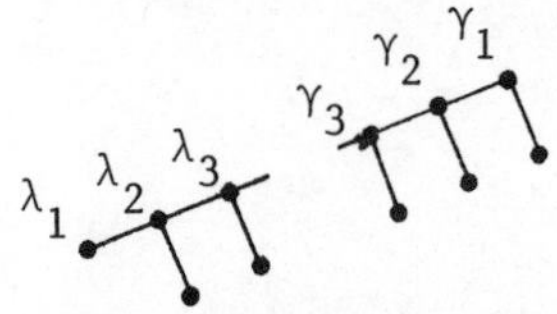

and let $G = \Sigma(\Gamma,R)$. Then $\cup H_{\overline{\lambda_i}} = \cap H_{\overline{\gamma_i}}$ belongs to $\mathcal{K}$ but is not an $H_{\overline{\alpha}}$ or an $H^{\overline{\alpha}}$.

EXAMPLE 6.4. Let $\Gamma = \{(x,0)$ and $(x,1) \mid x$ is real and $0 \leq x \leq 1\}$ with the partial order $(x,1) > (x,0)$. Let G be the real valued functions that are continuous on the top line and finitely non-zero on the bottom.

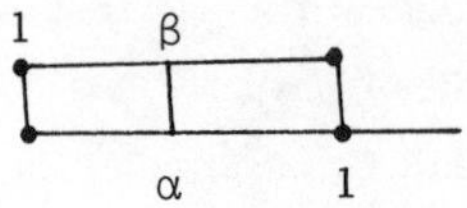

$G^{\beta} = G$ so $G^{\beta} \cap S_{\beta}' = S_{\beta}'$ consists of those functions which live on α and hence is a non-zero o-subgroup. But S_{β} is not maximal. In fact G_{β} contains an infinite number of minimal primes besides G_{α}.

EXAMPLE 6.5. Let $\Gamma =$ 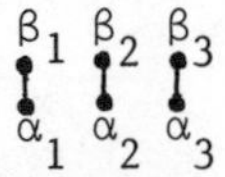

.... and let G consist of all the real valued functions that are periodic on the top and finitely non-zero on the bottom. Once again, G is an ℓ-subgroup of $V(\Gamma,R)$. Since the top group is hyperarchimedian, it follows that the primes of G are of the form G_{α_1} or G_{β_1} or those that are both maximal and minimal.

$K_j = \cap_{i \neq j} G_{\alpha_i}$ consists of those functions that live on α_j so $G_{\beta_j} = G_{\alpha_j} \boxplus K_j \in \mathcal{P}$. Thus all the primes belong to $\mathcal{P}$ so $\mathcal{P} = \mathcal{C}$. Since $\Gamma \neq \overline{\Gamma}$, $\mathcal{K}^* \subset \mathcal{P} = \mathcal{C}$. Now $C \in \underset{\sim}{B}$ so $\mathcal{K}$ consists of all the minimal primes together with G and hence $C_{\beta_j} \notin \mathcal{K}^*$. But G_{α_j} and K_j belong to $\mathcal{K}^*$ so $\mathcal{K}^*$ is not a sublattice of $\mathcal{C}$.

Consider $0 < g = \begin{pmatrix} b_1, b_2, \ldots \\ a_1, a_2, \ldots \end{pmatrix} \in G$. If $b_j = 0$ and $a_j \neq 0$ let K_j be as above. Note there are only a finite number of them, say $n_1,..,n_t$. Let $K = \cap\, \delta_{\alpha_i}$ for all i such that $b_i = 0$. Then $G(g) = K \boxplus K_{n_1} \boxplus \ldots \boxplus K_{n_t}$ and hence it follows that each element of $\mathcal{C}$ is the join of intersections of minimal primes.

Finally note that each S_{β_i} is maximal and β_i is essential but not special.

EXAMPLE 6.6. Let G be as in Example 5 and let $H = V(\Delta,R)$ where $\Delta =$. Then for $G \boxplus H$ we have $\mathcal{K}^* \subset \mathcal{P} \subset \mathcal{C}$.

REFERENCES

1. S. Bernau, The lateral completion of an arbitrary lattice group. *J. Australian Math. Soc. 19:* 263-289 (1975).

2. A. Bigard, P. Conrad, S. Wolfenstein, Compactly generated lattice-ordered groups. *Math. Zeitschr. 107:* 201-211 (1968).

3. A. Bigard, K. Keimel and S. Wolfenstein, *Groups et Anneaux Reticules*, Springer-Verlag, Berlin, 1977.

4. R. Byrd, *Lattice-Ordered Groups*, Thesis, Tulane University, 1966.

5. R. Byrd, Complete distributivity in lattice-ordered groups. *Pacific J. Math. 20:* 423-432 (1967).

6. R. Byrd and T. Lloyd, Closed subgroups and complete distributivity in lattice-ordered groups. *Math. Zeitschr. 101:* 123-130 (1967).

7. P. Conrad, J. Harvey and C. Holland, The Hahn Embedding theorem for abelian lattice-ordered groups. *Trans. Amer. Math. Soc. 108:* 143-169 (1963).

8. P. Conrad, The lattice of all convex ℓ-subgroups of a lattice-ordered group. *Czech. Math. J. 15:* 101-123 (1965).

9. P. Conrad, Representations of partially ordered abelian groups as groups of real valued functions. *Acta. Math. 116:* 199-221 (1966).

10. P. Conrad, Lex-subgroups of lattice-ordered groups. *Czech. Math. J. 18:* 86-103 (1968).

11. P. Conrad and D. McAlister, The completion of a lattice-ordered group. *J. Australian Math. Soc. 9:* 182-208 (1969).

12. P. Conrad, The lateral completion of a lattice-ordered group. *Proc. London Math. Soc. 19:* 444-486 (1969).

13. P. Conrad, *Lattice ordered groups*, Tulane University, 1970.

14. P. Conrad, The topological completion and the linearly compact hull of of an abelian ℓ-group. *Proc. London Math. Soc. 28:* 457-482 (1974).

15. P. Conrad, Torsion radicals of lattice-ordered groups. *Symposia Math. 21:* 479-513 (1977).

16. F. Sik, Zur theorie der halbgeordneten Gruppen. *Czech. Math. J. 10:* 400-424 (1960).

LEXICOGRAPHIC CENTERS OF A LATTICE-ORDERED GROUP

Jorge Martinez
University of Florida
Gainesville, Florida

1. INTRODUCTION

If C is a convex ℓ-subgroup of the lattice-ordered group G, the lex-center K(C) of C is defined as the subgroup generated by $\{0 \leq g \in C | g < u$ for all $0 < u \in G \setminus C\}$. K(C) is a closed convex ℓ-subgroup of G. If $C \neq 0$, $C = K(C)$ if and only if G is a lexicographic extension of C. Suppose K is a non-polar convex ℓ-subgroup; then K is a lex-center if and only if K″ is a lexicographic extension of K. K is then the lex-center of $K \boxplus K'$ and no other convex ℓ-subgroup.

Using the lex-center, the concept of a lexicographic extension is generalized to that of a *-lexicographic extension, via a series of convex ℓ-subgroups. Roughly speaking, G is a *-lexicographic extension of the convex ℓ-subgroup C if G can be built up using a system of convex ℓ-subgroups of C, by alternately taking lex-centers and polar-closures. In particular, if the special values of G form a plenary set reasonable sufficient conditions imply that G is a *-lexicographic extension of C.

All lattice-ordered groups (henceforth: ℓ-groups) shall be written additively unless otherwise noted. G^+ will denote the positive cone of an ℓ-group G; that is, $g \in G^+$ if and only if $g \geq 0$. $\mathscr{C}(G)$ shall stand for the lattice of convex ℓ-subgroups of the ℓ-group G. It is probably wise to list a few pertinent definitions here; see [1] for details. $M \in \mathscr{C}(G)$ is a value of G if it is maximal with respect to missing an element $g \in G$. If M* denotes the meet of all the convex ℓ-subgroups which properly contain M, then $g \in M^* \setminus M$ and no convex ℓ-subgroup of G lies strictly between M and M*. M is also said to be a *value* of g; M* is its cover. We say that an element $s \in G$ is *special* if it has only one value. The value of a special element is also called a *special value*. A convex ℓ-subgroup N of G is *prime* if $a \wedge b = 0$ implies that $a \in N$ or $b \in N$; it is well-known that every value is prime.

$P \in \mathcal{C}(G)$ is a *polar subgroup* if there is a subset X of G^+ such that $P = X' \equiv \{g \in G \mid |g| \wedge x = 0, \text{ all } x \in X\}$. (Note: $|g| = g \vee -g$.) Alternately, P is a polar if $P = P''$. A convex ℓ-subgroup C of G is *dense* in G if $C'' = G$; that is, if for each $0 < g \in G$ there exists an element $c \in C$ such that $0 < c \leq g$.

To conclude our preliminary definitions, recall that $C \in \mathcal{C}(G)$ is *closed* if C is closed under all existing suprema and infima of elements of C. Equivalently, C is closed if and only if for each family $\{c_\lambda \mid \lambda \in \Lambda\} \subseteq C^+$ for which $c = \bigvee c_\lambda$ exists, it follows that $c \in C$.

For each $C \in \mathcal{C}(G)$ let

$$I(C) = \bigcap\{S \in \mathcal{C}(G) \mid S \nleq C\}.$$

Let us observe some basic properties of I(C). First, if $S \in \mathcal{C}(G)$, then either $S \leq C$ or else $I(C) \leq S$. We also have the following unpublished proposition due to Conrad. Recall that a prime E of G is *essential* if there is an element $0 \neq g \in G$ having all its values beneath E. (Note: since E must contain a value to be essential, and values are prime, then E is necessarily prime (see [1]).)

PROPOSITION 1. Suppose $C \in \mathcal{C}(G)$. $I(C) = \{g \in G \mid \text{each value of } g \text{ is contained in } C\}$. Thus $I(C) \neq 0$ if and only if C is essential. C is a special value of G if and only if $I(C) \nleq C$.

Proof. Suppose $0 < g \in I(C)$ and M is a value of g. Since $I(C) \nleq M$ it follows that $M \leq C$. Conversely, suppose each value of g is contained in C. If $S \in \mathcal{C}(G)$ and $S \nleq C$ then $g \in S$. By definition of I(C), $g \in I(C)$. It is immediate now that $I(C) \neq 0$ if and only if C is essential.

If C is special and the only value of $0 < s \in G$ then it is clear that $I(C) = G(s)$, the convex ℓ-subgroup generated by s. Also $I(C) \cap C$ is the value of s in G(s). Evidently, $I(C) \nleq C$.

Conversely, if $I(C) \nleq C$, then we shall prove that I(C) is compact in the lattice of convex ℓ-subgroups. (Note: an element c in a lattice L is *compact* if $c \leq \bigvee x_i$ implies that $c \leq x_{i_1} \vee \cdots \vee x_{i_n}$ for a suitable choice of finitely many indices $i_1, \ldots, i_n \in I$. The compact elements of $\mathcal{C}(G)$ are precisely the principal convex ℓ-subgroups. This result can be found in [7].)

Suppose $I(C) \leq \bigvee A_\lambda (\lambda \in \Lambda)$, where each $A_\lambda \in \mathcal{C}(G)$, and no A_λ contains $I(C)$. Then each $A_\lambda \leq C$ and hence $I(C) \leq C$, a contradiction. Therefore, $I(C) \leq A_\mu$ for some $\mu \in \Lambda$, and in particular, $I(C)$ is compact. Set $I(C) = G(k)$ with $k > 0$. It is then straightforward to verify that C is the only value of k.

2. LEXICOGRAPHIC CENTERS OF AN ℓ-GROUP

Suppose $C \in \mathcal{C}(G)$; the *lex-center* of C is the subgroup K(C) generated by $\{g \in C^+ | g < u \text{ for all } u \in G^+ \setminus C\}$. It is easy to show that K(C) is a closed, convex ℓ-subgroup of G. Notice also that if $K(C) \neq 0$ then C is a closed, prime subgroup which is dense in G. In addition, if $C \neq 0$ then $C = K(C)$ if and only if G is a *lexicographic extension* of C; that is, C is prime and if $0 < u \in G \setminus C$ then $c < u$ for each $c \in C^+$. (Following Conrad in [3], we use the notation $G = \text{lex}(C)$ when G is a lexicographic extension of C).

The following lemma describes a lex-center in the language of o-permutation groups. In this lemma and in the proof of Proposition 3, we must use multiplicative notation.

LEMMA 2. Suppose the ℓ-group G is represented as a group of o-permutations on a chain Ω, and $C \in \mathcal{C}(G)$. Then $K(C) = \{g \in C | \alpha \in \Omega \text{ and } \alpha h = \alpha u$, for some $h \in C^+$, $u \in G^+ \setminus C$ imply that $\alpha g = \alpha\}$.

Proof. Let $K_0(C) = \{g \in C | \alpha \in \Omega$ and $\alpha h = \alpha u$, for some $h \in C^+$ and $u \in G^+ \setminus C$ imply that $\alpha g = \alpha\}$. Clearly $K_0(C)$ is a convex ℓ-subgroup of G. To show then that $K(C) = K_0(C)$ it suffices to prove that their positive cones agree.

Suppose $1 < g \in C \setminus K(C)$; then an element $u \in G^+ \setminus C$ exists such that $g \nleq u$. Pick $\alpha_0 \in \Omega$ such that $\alpha_0 g > \alpha_0 u$; then $h = g \wedge u$ belongs to C and $\alpha_0 h = \alpha_0 u$. Moreover $\alpha_0 g > \alpha_0 u \geq \alpha_0$, and so $\alpha_0 g \neq \alpha_0$. Hence $g \notin K_0(C)$. On the other hand, suppose $1 \leq c \in K(C)$ and $\alpha h = \alpha u$, with $h \in C^+$, $u \in G^+ \setminus C$. Without loss of generality, (replacing h by $h \wedge u$, if necessary,) suppose $h \leq u$. Then $uh^{-1} \in G^+ \setminus C$ and so $c \leq uh^{-1}$; that is, $ch \leq u$. In particular, $\alpha ch \leq \alpha u = \alpha h$, which implies $\alpha c \leq \alpha$. Since $c \geq 1$ we conclude that $\alpha = \alpha c$, proving that $c \in K_0(C)$.

This completes the proof of the lemma.

PROPOSITION 3. For any convex ℓ-subgroup C of G, $I(C) \cap C = K(C)$.

Proof. Lemma 2 guarantees that $C \cap I(C) \leq K(C)$: Suppose G is represented on a chain Ω as before. If $1 \leq c \in C \cap I(C)$ and $\alpha \in \Omega$ with $\alpha h = \alpha u$, where $h \in C^+$ and $u \in G^+ \setminus C$ and $h < u$, then the stabilizer G_α contains uh^{-1}. In addition, since $uh^{-1} \notin C$, $G(uh^{-1}) \leq C$, which implies that $I(C) \leq G(uh^{-1})$ and $G(uh^{-1}) \leq G_\alpha$. Consequently $c \in G_\alpha$; ie. $\alpha c = \alpha$ and $c \in K(C)$.

As for the reverse containment, the definition of K(C) implies that K(C) also has the property: if $S \in \mathcal{C}(G)$ then either $S \leq C$ or $K(C) \leq S$. I(C) is obviously the *largest* convex ℓ-subgroup of G with this property relative to C. Hence, $K(C) \leq I(C)$ and so $K(C) = C \cap I(C)$.

COROLLARY. For $C \in \mathcal{C}(G)$, $K(C) = \{g \in C \mid \text{each value of } g \text{ is properly contained in } C\}$.

COROLLARY. For $C \in \mathcal{C}(G)$, $K(C) \neq 0$ if and only if C is a dense, essential subgroup.

COROLLARY. Lex-centers can be distinguished (lattice-theoretically) in $\mathcal{C}(G)$.

Our next proposition collects some basic properties of lex-centers.

PROPOSITION 4. Suppose $C \in \mathcal{C}(G)$. Then

(i) if $K(C) \neq 0$, $K(C)' \leq C$.

(ii) $K(C)'' \cap C = K(C)$ and $I(C) \leq K(C)''$.

(iii) If C is normal in G so is K(C). Then $G/K(C)' = \text{lex}(C/K(C)')$.

(iv) Computed in $G/K(C)$, $K(C/K(C)) = 0$.

Proof. (i) Suppose $0 < x \in K(C)'$ and select $0 < c \in K(C)$. If $x \notin C$ then $c < x$, which contradicts the disjointness of c and x.

(ii) We suppose $K(C) \neq 0$, otherwise there is nothing to prove. Suppose $0 \leq c \in C$ and $u \in G^+ \setminus C$; look at $(-u + c) \vee 0$. Now $c \wedge u \in C$ and therefore $(c \wedge u) + g < u$, for all $g \in K(C)^+$. So $c \wedge (u + g) \leq (c + g) \wedge (u + g) = (c \wedge u) + g < u$, which implies that $(-u + c) \wedge g \leq 0$, and hence that $[(-u + c) \wedge g] \vee 0 = 0$. Putting it differently, for each $g \in K(C)^+$

$$g \wedge [(-u + c) \vee 0] = 0.$$

Hence: $(-u + c) \vee 0 \in K(C)'$. Thus, if $c \in K(C)'' \cap C$ then $c \wedge [(-u + c) \vee 0] = 0$. Next, $(u + c) \wedge (c \vee u) = u$, which says that $c \vee u = u$; ie. $c \leq u$, and therefore $c \in K(C)$. Conclusion: $K(C)'' \cap C = K(C)$.

The inequality follows from Proposition 3 and the fact that $K(C)''$ is the largest convex ℓ-subgroup satisfying the equation $X \cap C = K(C)$.

(iii) and (iv) are obvious.

In general $K(C)$ need not be a cardinal summand of C. Consider for example:

$$G = \{((a_n),(b_n),c) \mid (a_n),(b_n) \text{ are eventually constant sequences of integers, and } c \in Z.\}$$

Addition in G is coordinate-wise. We set $((a_n),(b_n),c) \geq 0$ if each $(a_n) \geq 0$ and if $a_m = 0$ then $b_m \geq 0$; if (a_n) and (b_n) are both finitely non-zero, we require that $c \geq 0$.

Let $C = \{((a_n),(b_n),c) \in G \mid (a_n) \text{ is finitely non-zero}\}$. Then $K(C) = \{(0,0,c) \mid c \in Z\}$, while $K(C)' = \{((a_n),(b_n),0) \in G \mid$ *both* (a_n) and (b_n) are finitely non-zero$\}$. Clearly: $C \neq K(C) + K(C)'$.

The next result gives a sufficient condition for splitting a convex ℓ-subgroup C as a cardinal sum of $K(C)$ and its polar.

THEOREM 5. Suppose $C \in \mathscr{C}(G)$ and $K = K(C)$. If $K \neq K''$ then $C = K \boxplus K'$.

Proof. Our assumption implies that $K \neq 0$ and $K'' \nleq C$, in view of Proposition 4. Next, observe that if $0 \leq g \in C$ and $0 < a, b \in K'' \setminus C$, then $g \wedge a = g \wedge b$. For $g \wedge b \in K$ and therefore $g \wedge b < a$ whence $g \wedge b \leq g \wedge a$; if we reverse the roles of a and b it follows that $g \wedge b = g \wedge a$. Thus the set $\{g \wedge a \mid 0 < a \in K'' \setminus C\}$ is a singleton for each $g \in C$; denote it by x_g. The crucial claim is that $y = g - x_g \in K'$: for $y \in C$, and the set $\{y \wedge a \mid 0 < a \in K'' \setminus C\} + x_g = \{(y + x_g) \wedge (a + x_g) \mid 0 < a \in K'' \setminus C\} = \{g \wedge a \mid 0 < a \in K'' \setminus C\} = \{x_g\}$. The upshot of all this is that $x_y = 0$; that is, $y \wedge a = 0$ for all $0 < a \in K'' \setminus C$. In particular: $y \in K'$.

We are now almost able to characterize lex-centers within $\mathscr{C}(G)$.

THEOREM 6. Suppose $K \in \mathscr{C}(G)$ and $K \neq K''$. Then K is a lex-center if and only if $K'' = \mathrm{lex}(K)$. If this is the case then K is the lex-center of $K \boxplus K'$ and no other convex ℓ-subgroup.

Proof: If $K = K(C)$ for a suitable convex ℓ-subgroup C then by Theorem 5 $C = K \boxplus K'$ and is therefore uniquely determined. Moreover, since $K'' \cap C = K$, $K'' = \text{lex}(K)$.

Conversely, suppose $K'' = \text{lex}(K)$ and set $C = K \boxplus K'$. By Clifford's Lemma (see [3]) $K < u$ $(u \in G^+)$ if and only if $u \notin C$. Immediately $K \leq K(C)$. On the other hand, if $g \in C^+$ and $g < u$ for all $u \in G^+ \setminus C$ then the same is true for all $0 < u \in K'' \setminus K$. Hence $g \in K$ and $K = K(C)$.

Recall (from [3]) that $L \in \mathscr{C}(G)$ is a *lex-subgroup* if there is a proper convex ℓ-subgroup U of L so that $L = \text{lex}(U)$. We can summarize our last theorem by saying that non-polar lex-centers are to be located between maximal lex-subgroups and their lex-kernels.

To recall the notion of a *lex-kernel* we shall present for the reader's reference two theorems from [3]. An element $0 < u \in G$ is a *unit* if $u \wedge g = 0$ implies that $g = 0$. The lex-kernel of G is the subgroup of G generated by the *non*-units of G.

THEOREM 7. (Conrad [3], Lavis [6]) Let N denote the set of non-units of G. Then

$$\begin{aligned} \langle N \rangle &= \bigvee \{M \in \mathscr{C}(G) \mid M \text{ is a minimal prime}\} \\ &= \bigvee \{a'' \mid a'' \neq G\} \end{aligned}$$

THEOREM 8. (Conrad [3]) For $0 \neq C \in \mathscr{C}(G)$ the following are equivalent.

(a) $G = \text{lex}(C)$.

(b) $C \geq \langle N \rangle$.

(c) C is comparable to every convex ℓ-subgroup of G.

To capsule the second theorem: the lex-kernel is the smallest convex ℓ-subgroup which is comparable to every other convex ℓ-subgroup.

A lex-center is a bounded subgroup. ($B \in \mathscr{C}(G)$ is *bounded* if $B < x$ for some $x \in G^+ \setminus B$.) We give an example in which the only lex-centers are trivial, but which has non-zero bounded subgroups.

Let $G = \{((a_n), (b_n)) \mid (a_n)$ and (b_n) are periodic sequences of integers, and the period of (b_n) divides that of $(a_n)\}$. G is ordered by $((a_n), (b_n)) \geq 0$ if each $b_n \geq 0$ and wherever $b_m = 0$ then $a_m \geq 0$. It can be verified that $K(C) = 0$ for each proper convex ℓ-subgroup C of G. Yet $A = \{((a_n), 0) \mid (a_n)$ periodic$\}$ is a bounded convex ℓ-subgroup.

One can modify G slightly, and obtain an ℓ-group in which two distinct primes have the same *non-trivial* lex-center.

We shall conclude this section with two results about the structure of the set of lex-centers. The proofs are straightforward and will be left to the reader.

PROPOSITION 9. If $\{C_\lambda | \lambda \in \Lambda\}$ is a family of convex ℓ-subgroups of G then $K(\cap C_\lambda) = \cap K(C_\lambda)$.

PROPOSITION 10. If $C,D \in \mathscr{C}(G)$ and $C||D$ then $K(C) \cap K(D) = 0$.

In particular, the set of lex-centers of G forms a lattice. The proper, non-zero lex-centers form a root-system which is a meet-homomorphic image of the set of essential primes.

3. VARIOUS LEXICOGRAPHIC SERIES

In what follows we assume that C is a fixed convex ℓ-subgroup of the ℓ-group G. A *lex-series* for C is a family $\{(T_\lambda, T^\lambda) | \lambda \in \Lambda\}$ of pairs of convex ℓ-subgroups of C, where Λ is a p.o. set, each $T^\lambda \neq 0$, and

(1) $T_\lambda \leq T^\lambda$ for each $\lambda \in \Lambda$,

(2) $\lambda < \mu$ implies that $T^\lambda \leq T_\mu$, whereas $\lambda || \mu$ implies that $T_\lambda \cap T_\mu = 0$,

(3) $T^\lambda = C \cap K(C \vee T_\lambda'')$ for each $\lambda \in \Lambda$, and

(4) $G = [\cup T^\lambda]''$ while $\cap T_\lambda = 0$.

If a lex-series exists for C we say that G is a **-lex-extension* of C and write G = *-lex(C). It is evident that if G = lex(C) then G = *-lex(C), using the lex-series $\{(0,C)\}$.

As examples of lex-series we give two contrasting cases. Consider first the V-group on the following root-system Δ_1:

Δ_1:

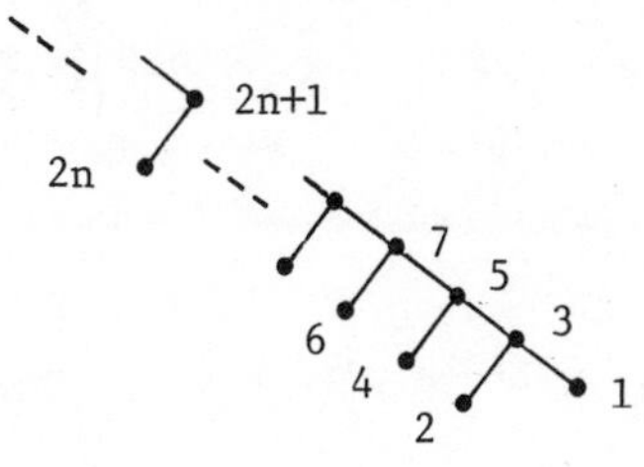

Thus $G = V(\Delta_1, \mathbb{R})$; let $C = \{g \in G | g_{2n+1} = 0,\ n = 1,2,\ldots\}$. Then G = *-lex(C); we shall produce a lex-series for C. Let $T_k = \{g \in G | g \in C$ and $g_n = 0$ for

$n > 2k\}$; let $T^k = T_{k+1}$ and $T_0 = 0$, $T^0 = T_1$. Then $\{(T_k,T^k) | k = 0,1,2,\ldots\}$ is a lex series for C.

Next we look at the V-group on the root-system Δ_2 below:

Δ_2:

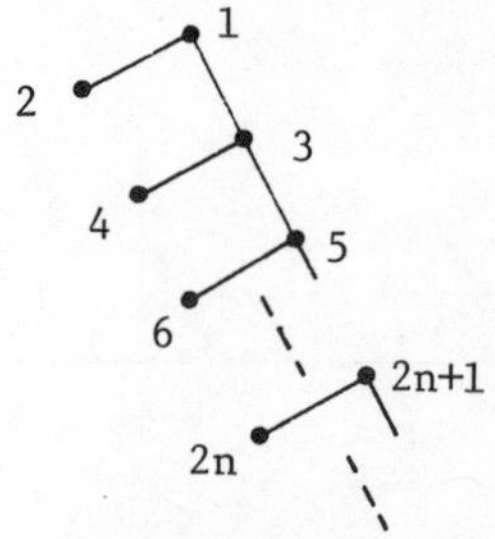

Here $G = V(\Delta_2, \mathbb{R})$ and $C = \{g \in G | g_{2n+1} = 0, n = 0,1,2,\ldots\}$. Once again $G =$ *-lex(C), this time with a descending lex-series. For each natural number n let $T^n = \{g \in C | g_k = 0, k < 2n\}$; set $T_n = T^{n+1}$. Then $\{(T_n,T^n) | n = 1,2,\ldots\}$ is a lex-series for C.

These two examples are not atypical. If $\{(T_\lambda,T^\lambda) | \lambda \in \Lambda\}$ is any lex-series and $\lambda || \mu$ we have by (2) that $T_\lambda \cap T_\mu = 0$. This implies that $T_\lambda'' \cap T_\mu'' = 0$ and hence that $C = (C \vee T_\lambda'') \cap (C \vee T_\mu'')$. By Proposition 9, $K(C) = K(C \vee T_\lambda'') \cap K(C \vee T_\mu'')$, from which we conclude that $K(C) = K(C \vee T_\lambda'')$ or $K(C) = K(C \vee T_\mu'')$. In particular, the set $\{T^\lambda | \lambda \in \Lambda\}$ is a chain under containment. Putting it differently, if $\lambda || \mu$ in Λ then *both* λ and μ are minimal, and either $T^\lambda = K(C)$ or $T^\mu = K(C)$. If $K(C) = 0$ then Λ is a chain.

Suppose $K(C) \neq 0$. We may then replace all pairs (T_λ,T^λ) with $T^\lambda = K(C)$ by the single pair $(0,K(C))$. This process is called the *simplification* of a lex-series, and we shall say (after the change) that it is in *simplified* form. If C has non-zero lex-center and has a lex-series we shall always assume that the lex-series is in simplified form.

Suppose once again that $\{(T_\lambda,T^\lambda) | \lambda \in \Lambda\}$ is a lex-series for C and that $K(C) \neq 0$. If $\gamma \in \Lambda$ exists such that $T_\gamma \geq K(C)$ then by our earlier arguments, Λ is a chain. We conclude then that the only way Λ can fail to be a chain is for the lex-series to consist of two pairs $\{(0,K(C)), (T_\mu,C)\}$, with C dense in G. We shall refer to it as a *singular* lex-series. Presently we shall see that if C has a singular lex-series then $G = \text{lex}\,(C \vee T_\mu'')$, or else $G = C \vee T_\mu''$.

First we need a technical result about lex-series.

PROPOSITION 11. Suppose $C \in \mathscr{C}(G)$ and $\{(T_\lambda,T^\lambda) | \lambda \in \Lambda\}$ is a lex-series for C. If C is dense in G we have for each $\lambda \in \Lambda$,

(i) $T^\lambda = C \cap (T^\lambda)''$ and $K(C \vee T_\lambda'') \leq (T^\lambda)''$.

(ii) $(T^\lambda)'' = K(C \vee T_\lambda'')''$.

(iii) $T^\lambda \vee T_\lambda'' = K(C \vee T_\lambda'')$.

Proof. Now $(T^\lambda)'' = C'' \cap K(C \vee T_\lambda'')'' = G \cap K(C \vee T_\lambda'')'' = K(C \vee T_\lambda'')''$, proving (ii). Further: $C \cap (T^\lambda)'' = C \cap K(C \vee T_\lambda'')'' = C \cap (C \vee T_\lambda'') \cap K(C \vee T_\lambda'')'' = C \cap K(C \vee T_\lambda'') = T^\lambda$, which is half of (i). The other half is a consequence of properties of polars. We leave (iii) to the reader.

COROLLARY. If C has a lex-series $\{(T_\lambda, T^\lambda) | \lambda \in \Lambda\}$ and Λ is not a chain then $K(C) \neq 0$, C is dense in G and the lex-series is singular. Furthermore C has a singular lex-series if and only if $K(C) \neq 0$. C is dense in G and there exists a $T \in \mathcal{C}(G)$ such that $T || C$, $T \vee C = G$ or else $G = \text{lex } (T \vee C)$.

Proof. By our discussion preceding Proposition 11, the lex-series consists of two pairs $\{(0,K(C)), (T_\mu, T^\mu)\}$ if Λ is not a chain, with $T^\mu \geq K(C)$. Now $(T^\mu)'' = G$ which by Proposition 11 means that $T^\mu = C$ and $C \vee T_\mu'' = K(C \vee T_\mu'')$. The rest is clear.

Call a convex ℓ-subgroup *singular* if it has a singular lex-series. Neither of the two examples described at the beginning of this section are singular.

There is a canonical series which satisfies (1) through (3) in the definition of lex-series. We shall give some reasonable conditions for (4) to hold. In preparation for this discussion we must first discuss the notion of root-equivalence; we refer the reader to [4] for details.

Two primes in an ℓ-group G, say M and N, are said to be *root-equivalent* if they contain the same minimal primes. (Alternately: if they lie on the same maximal chains of the root-system of primes.) We shall denote the set of root-equivalence classes by Γ. If $\gamma \in \Gamma$ we set

$$H_\gamma = \cap \{M | M \in \Gamma\} \text{ and } H^\gamma = \cup \{M | M \in \Gamma\}.$$

Put $\gamma_1 < \gamma_2$ if $H^{\gamma_1} \leq H_{\gamma_2}$. Γ then also becomes a root-system.

We quote from Conrad [4] in the next three propositions.

PROPOSITION 12. $\{H_\gamma | \gamma \in \Gamma\}$ is the set of meet-irreducible elements among the primes which are joins of minimal primes.

PROPOSITION 13. G is representable (as a subdirect product of o-groups) if and only if each H_γ is normal in G.

PROPOSITION 14. G is normal valued (that is, M is normal in M* for each value M) if and only if each H_γ is normal in H^γ.

PROPOSITION 15. Suppose M is a special value of G and $H_\gamma \leq M < H^\gamma$. Suppose $0 < a \in G$ is special and $G(a) = I(M)$. Then

(i) $H_\gamma \cap a'' = H_\gamma \cap G(a)$ is the lex-kernel of a''.

(ii) For each prime N such that $H_\gamma \leq N < H^\gamma$, $K(N) = N \cap a''$ and $N = K(N) \boxplus a'$.

(iii) Each value in γ is special.

Proof. From Theorem 3.8 in [3], G(a) is a lex-subgroup and a'' the maximal lex-subgroup containing it. Let Q denote the lex-kernel of a''. We claim first that $H^\gamma = a'' \boxplus a'$. For $a' \leq H_\gamma < H^\gamma$ (since H_γ is prime and $a \notin H_\gamma$). Evidently then $a'' \boxplus a' \leq H^\gamma$. If $a'' \boxplus a' < H^\gamma$ and $0 < b \in H^\gamma \setminus (a'' \boxplus a')$ then by Clifford's Lemma (see [3]) $b > a''$. In particular, $a'' < G(b)$, Moreover, $b' \leq a' \leq H^\gamma$, and so $G(b) \boxplus b' \leq H^\gamma$. That makes $G(b) \boxplus b'$ the cover of a value of b lying in γ. Hence it is the cover of all the values of b, and since $G(b) \boxplus b'$ is root-equivalent to $M \in \gamma$, it follows that b is special. But then by Theorem 3.8 of [3] once again, $b'' = a''$, which contradicts the preceding arguments. Hence $H^\gamma = a'' \boxplus a'$ as claimed.

The map $N \to N \cap a''$ is therefore a one-to-one correspondence between the primes of the root-equivalent class γ of M and all the primes of a'' over Q. (The inverse map is $P \to P \boxplus a'$.) The statements in (i) and (iii) are then clear.

Now for (ii): if $H_\gamma \leq N < H^\gamma$ then $a'' = \mathrm{lex}(N \cap a'')$, from which it easily follows that $K(N) \leq N \cap a''$. Since $N \cap a''$ is not a polar, and by Theorem 5: $N = K(N) \boxplus a'$. This implies that $K(N) = N \cap a''$ since cardinal complements are unique in the lattice of convex ℓ-subgroups.

Once again suppose that $C \in \mathscr{C}(G)$ and is dense in G. Define a subset Δ of Γ (the set of root-equivalence classes) by: $\delta \in \Delta$ if (i) $H^\delta \geq C$ and (ii) δ consists of special subgroups. (Note: by Proposition 15 either no values between H_δ and H^δ are special or else they all are.)

For each $\delta \in \Delta$ define: $K^\delta = C \cap K(a'' \boxplus a')$, where a is a special element so that $H^\delta = a'' \boxplus a'$; $K_\delta = C \cap K(H_\delta)$. It is then easy to verify that the family $\{(K_\delta, K^\delta) \mid \delta \in \Delta\}$ satisfies conditions (1) through (3) in the definition of a lex-series. In addition, since C is dense in G, each $K^\delta \neq 0$.

Before stating the next theorem we define $C_{card} = \cap\{a' \mid a \text{ is special and } a \notin C\}$.

THEOREM 16. Suppose C is a dense convex ℓ-subgroup of G and (a) $C_{card} = 0$ while (b) $\cap\{H_\delta \mid \delta \in \Delta\} \leq C$, then $\{(K_\delta, K^\delta) \mid \delta \in \Delta\}$ or its simplification is a lex-series and G = *-lex(C).

Proof. We prove first that $\cap K_\delta = 0$. Now, $\cap K_\delta = C \cap (\cap K(H_\delta)) \leq C \cap K(C) = K(C)$. Thus if $K(C) = 0$, $\cap K_\delta = 0$; otherwise we pass to the simplification of our lex-series to handle this condition.

To show $G = [\cup K^\delta]''$ we must prove that $\cap (K^\delta)' = 0$. (Note: if some $C \vee K_\delta'' = G$, or $C = K^\delta$, there is nothing to prove.) Now $(K^\delta)'' = K(a'' ⊞ a')'' \geq a''$, and so $a' \geq (K^\delta)'$. Since $C_{card} = 0$, $\cap (K^\delta)' = 0$ as well.

Conclusion: the series $\{(K_\delta, K^\delta) \mid \delta \in \Delta\}$ or its simplification is a lex-series. Hence: G = *-lex(C).

Recall that a set $\mathcal{M}$ of values of an ℓ-group G is said to be *plenary* if (i) $\cap \mathcal{M} = 0$ and (ii) for each $M \in \mathcal{M}$ and each value N of G, $N \geq M$ implies $N \in \mathcal{M}$. For a detailed discussion of the role of plenary subsets see [5].

COROLLARY. Suppose the special values of G form a plenary set. Suppose that $C \in \mathcal{C}(G)$ is closed and dense. Then if either $C_{card} = 0$ or some $H^\delta = G$ $(\delta \in \Delta)$, G = *-lex(C).

Proof. To apply Theorem 16 we need to show that $\cap \{H_\delta \mid \delta \in \Delta\} \leq C$. Recall from [2] that the following condition is equivalent to our hypothesis that the special values of G form a plenary set: each $0 < g \in G$ can be written as the join (possibly infinite) of pairwise disjoint, special elements. Thus, if $0 < g \notin C$, then since C is closed, there is a special element $0 < s \notin C$ such that $s \leq g$ and its value M coincides with a value of g. In particular, $M \geq C$ and so we can conclude that $\cap H_\delta \leq C$.

We make a brief aside here to discuss the role of C_{card}. From the discussion in [2] it follows that every special value is closed. Assuming that C is closed and that the special values of G form a plenary set we obtain, using the characterization mentioned in the proof of the preceding corollary, that $C_{card} \leq C$. Moreover, it is easy to verify that a special element is either in C_{card} or in C_{card}'. Hence: $G = C_{card} \boxplus C_{card}'$. Furthermore, if we set $C_{lex} = C_{card}'$, the reader can check that $C_{lex} = [\cup k^{\delta}]''$. We've therefore proved:

THEOREM 17. Suppose $C \in \mathscr{C}(G)$ is closed and dense in G. If the special values of G form a plenary set then $G = C_{card} \boxplus C_{lex}$ and $C_{lex} = *\text{-lex}(C \cap C_{lex})$.

We now turn to more special lex-series. A lex-series $\{(T_\lambda, T^\lambda) | \lambda \in \Lambda\}$ is *exact* if

(a) $T^\lambda = T_\mu$ whenever μ covers λ,

(b) $T^\lambda = \cap \{T_\mu | \mu > \lambda\}$ if $\lambda = \bigwedge\{\mu \in \Lambda | \mu > \lambda\}$,

(c) $T_\lambda = [\cup T^\mu]''$ $(\mu > \lambda)$, if $\lambda = \bigvee\{\mu \in \Lambda | \mu < \lambda\}$.

We say that G is a *transfinite lex-extension* of $C \in \mathscr{C}(G)$, written $G = T\text{-lex}(C)$, if C has an exact lex-series $\{(T_\lambda, T^\lambda) | \lambda \in \Lambda\}$, and Λ satisfies the DCC.

From our comments following the definition of lex-series we can conclude: if $\{(T_\lambda, T^\lambda) | \lambda \in \Lambda\}$ is exact and satisfies the DCC then $K(C) \neq 0$. For otherwise Λ is a chain, which is well-ordered, and therefore has a *least* element 0. By defining condition (4), $T_0 = 0$, making $T^0 = K(C)$, a contradiction.

We have proved the next result.

PROPOSITION 18. Suppose $G = T\text{-lex}(C)$. Then either C is singular or C has a well-ordered, exact lex-series $\{(T_\alpha, T_{\alpha+1}) | \alpha < \tau\}$ (τ is some ordinal) defined by:

(i) $T_1 = K(C)$;

(ii) $T_{\alpha+1} = C \cap K(C \vee T_\alpha'')$ for each $\alpha < \tau$;

(iii) $T_\beta = [\cup T_\alpha]''$ $(\alpha < \beta)$ if β is a limit ordinal.

From Theorem 16 we can extract the following sufficient conditions for G to be a transfinite lex-extension of C.

THEOREM 19. Suppose $C \in \mathscr{C}(G)$ is closed and dense in G. If

(i) each $\delta \in \Delta$ consists of special values, and

(ii) Δ satisfies the DCC,

then $G = T\text{-lex}(C)$, provided $C_{card} = 0$ or Δ has a maximal element. In the latter case C is singular.

Proof. The family $\{(K_\delta, K^\delta) | \delta \in \Delta\}$ is exact by (i), because if $\gamma < \delta$ and δ succeds γ in Δ then $H_\delta = H^\gamma$. The series satisfies the DCC by (ii). Now apply Theorem 16.

As a special case of Theorem 19, let us derive a theorem of Conrad from [3]. First some definitions: an ℓ-group G is a *lex-sum* of the o-groups $\{A_\lambda | \lambda \in \Lambda\}$ if there exist an ordinal σ and a chain $A^0 < A^1 < \ldots < A^\alpha < \ldots$ in $\mathscr{C}(G)$ such that

(a) $G = \cup A^\alpha$, and $A^\alpha = ⊞ A_\lambda^\alpha$ $(\lambda \in \Lambda_\alpha)$, where each A_λ^α admits no proper lex-extension and

(b) $\Lambda_0 = \Lambda$, $A_\lambda^0 = A_\lambda$ (all $\lambda \in \Lambda$).

(c) $A_\lambda^{\alpha+1} = A_\mu^\alpha$ for some $\mu \in \Lambda_\alpha$, or else $A_\lambda^{\alpha+1}$ is a proper lex-extension of a cardinal sum of two or more of the A_μ^α's. In addition, if $\beta > \alpha$ and A_λ^β is a lex-extension of $⊞ A_\delta^\alpha$ (some $\delta \in \Lambda_\alpha$), then $A_\lambda^\beta = A_\mu^{\alpha+1}$ for some $\mu \in \Lambda_{\alpha+1}$.

(d) If β is a limit ordinal then A_λ^β is a maximal lex-extension of the join of a chain of $A_{\lambda_\alpha}^\alpha$, one λ_α per $\alpha < \beta$.

Recall that an ℓ-group G is *finite-valued* if each non-zero element of G has at most finitely many values. Let M(G) denote the family of lex-subgroups of G.

THEOREM 20. (Conrad [3]) G is a lex-sum of o-groups if and only if G is finite valued and M(G) satisfies the DCC.

Proof. The necessity is easy; it is the sufficiency that interests us here. Suppose then that G is finite valued and that M(G) satisfies the DCC. The minimal lex-subgroups are then totally-ordered, and it is evident that G has a basis. Denote a basis for G by $\{G_i | i \in I\}$, and let B be the basic group of G. B is closed and dense in G, and the hypothesis for Theorem 19 and 17 are satisfied. Therefore $G = B_{card} ⊞ B_{lex}$ while $B_{lex} = \text{T-lex}(B \cap B_{lex})$.

To exhibit the lex-sum we use the canonical lex-series as follows: let $B_\alpha^0 = B$, $B^1 = B^0 \vee [\bigvee\{K_\delta'' | \delta \text{ is minimal in } \Delta\}]$. For any ordinal α, let $\Delta^\alpha = \{\delta \in \Delta | K_\alpha'' \leq B^\alpha\}$; define $B^{\alpha+1} = B^0 \vee [\bigvee\{K_\delta'' | \delta \text{ is the join of elements of } \cup\{\Delta^\alpha | \alpha < \beta\}\}]$.

G is then a lex-sum of the o-groups b_i'' $(i \in I)$.

4. SUMMARY

It seems fair to say that the major gap in the information supplied by this article is in Theorem 6: what do *polar* lex-centers look like? If they were all lex-subgroups then Theorem 16 ought to have a rather automatic converse. Unfortunately, polar lex-centers need not be lex-subgroups. Consider the following: let K be any ℓ-group having no special elements. Define G to be the ℓ-group of all pairs $((a_n),x)$, where (a_n) is an eventually constant sequence of integers and $x \in K$. The operation on G is coordinate-wise and we set $((a_n),x) \geq 0$ if each $a_n \geq 0$, and if (a_n) is finitely non-zero, then $x \geq 0$. We let $C = \{((a_n),x) \in G \mid (a_n)$ is finitely non-zero$\}$. Then $K(C) = K = K(C)''$, but K(C) has no special elements and cannot be a lex-subgroup. Incidently, it can be easily verified that $G \neq$ *-lex(C).

In view then of the fact that polar lex-centers are rather hard to characterize it seems probable that the conditions in Theorem 16 are a long way from being necessary for G to be a *-lex-extension of C.

REFERENCES

1. P. Conrad, The Lattice of all convex ℓ-subgroups of a lattice-ordered group. *Czech. Math J. 15 (90):* 101-123 (1965).

2. P. Conrad, A characterization of lattice-ordered groups by their convex ℓ-subgroups. *J. Austral. Math. Soc. 7:* 145-159 (1967).

3. P. Conrad, Lex-subgroups of lattice-ordered groups. *Czech. Math. J. 18 (93):* 86-103 (1968).

4. P. Conrad, The structure of an ℓ-group that is determined by its minimal prime subgroups, this volume.

5. P. Conrad, J. Harvey and C. Holland, The Hahn embedding theorem for abelian lattice-ordered groups. *Trans. Amer. Math. Soc. 108:* 143-169 (1963).

6. A. Lavis, Sur les quotients totalement ordonnes d'un group lineairement ordonne. *Bul. Soc. Royal Sci. Liege 32:* 204-208 (1963).

7. J. Martinez, Algebraic lattices. *Algebra Universalis 3 fasc. 2:* 247-260 (1973).

ARCHIMEDEAN LATTICE-ORDERED GROUPS WITH THE SPLITTING PROPERTY

G. Ya. Rotkovich and A. V. Koldunov

Hertzen Pedagogical Institute
Leningrad, U.S.S.R.

Lattice-ordered groups (ℓ-groups) with the splitting property have been examined by a number of mathematicians. The most interesting recent work is [1], in which some conditions are proved to be equivalent to the splitting property. Here we present some new results in this direction.

Terms having the same meaning for both ℓ-groups and for vector lattices are as in [2] and [3]. If the same concept is termed differently, however, then preference is given to [3] - except instead of "component", "band" is used. In what follows, only Archimedean ℓ-groups will be considered.

An ℓ-group X possesses the *splitting property* if it is a band with projections into any ℓ-group Y in which it is an ℓ-ideal.

LEMMA 1. For an ℓ-group X, the splitting property is equivalent to the condition that if X is a dense ℓ-ideal (foundation) in Y, then X = Y.

In an ℓ-group X, a class

$$\{x_\alpha \in X^+ | \alpha \in A\} \tag{I}$$

is called *semidisjoint* if for any $\alpha \in A$ the set $\{\beta \in A | x_\alpha \wedge x_\beta \neq 0\}$ is finite.

THEOREM 2. For an ℓ-group X, the splitting property is equivalent to the condition that if there is a semidisjoint set (I) such that for any $x \in X$, $\sup(x \wedge x_\alpha) \in X$, then the set (I) is bounded in X.

We will denote by $\mathbb{1}$ the order-unit in an ℓ-group X.

REMARK. A complete ℓ-group with the splitting property is laterally complete, i.e. any disjoint set has a supremum. Even for σ-complete ℓ-groups, this is not the case; namely there exists an example of a σ-complete ℓ-group without a $\mathbb{1}$ having the splitting property.

LEMMA 3. For an ℓ-group X with $\mathbb{1}$ and the splitting property, the condition that $\mathbb{1}$ is a strong order unit is equivalent to the condition that $x_n \downarrow 0$ implies $x_n \overset{r}{\downarrow} 0$ with regulator $\mathbb{1}$.

As an example of such a group, we can use $\mathscr{C}(\beta\mathbb{N}\setminus\mathbb{N})$.

A sequence $\{x_n \in X^+ | n \in \mathbb{N}\}$ is called a *tower* if $x_n \wedge m\mathbb{1} = x_m$ $(m \leq n)$ and $\{n\mathbb{1} - x_n | n \in \mathbb{N}\}^{\perp\perp} = X$.

THEOREM 4. For an ℓ-group X with $\mathbb{1}$, the splitting property is equivalent to the condition that a tower $\{x_n\}$ with $\sup(x \wedge x_n) \in X$ for any $x \in X^+$, must be bounded in X.

An ℓ-group X is called o-complete if from $|x_{n+p} - x_n| < v_n \downarrow 0$, it follows that there exists $x \in X$ such that $|x - x_n| < v_n \downarrow 0$.

The collection $\mathscr{D}_\infty(B)$ of all continuous extended (i.e. allowing values of $\pm\infty$ on a nowhere dense set) functions on a compactum B is a lattice, but not necessarily an ℓ-group. If $\mathscr{D}_\infty(B)$ is an ℓ-group (hence, a vector lattice), then it is o-complete. In this case, B will be called an o-*compactum*.

THEOREM 5. For an o-complete ℓ-group X with $\mathbb{1}$, the following statements are equivalent:

1) X has the splitting property.
2) Any tower in X has a supremum.
3) X is isomorphic to an ℓ-group X(B) of continuous extended functions on an o-compactum B. In this case, X(B) majorizes $\mathscr{D}_\infty(B)$, separates points in B and contains the constant function $\mathbb{1}$.

REMARK. If X is not o-complete, then the implication 3) → 1) is true, but 1) → 3) and 3) → 2) are not valid.

THEOREM 6. In an o-complete ℓ-group X with $\mathbb{1}$ in which for any $x \in X^+$ there exists $y \in X^+$ such that $2y \leq x \leq 4y$ (this condition is fulfilled in divisible ℓ-groups and therefore in vector lattices), the following conditions are equivalent:

1) X has the splitting property.

2) X is isomorphic to $\mathcal{D}_\infty(B)$ on an o-compactum B.

REMARK. In Theorems 5 and 6, it is impossible to replace the condition of o-completeness with the weaker condition of r-completeness, since under the Continuum Hypothesis there exists an r-complete ℓ-group $X \subseteq \mathcal{D}_\infty(\beta\mathbb{N})$ which has the splitting property and contains the constant function 1 but which has a tower unbounded in X.

REFERENCES

1. M. Anderson, P. Conrad, O. Kenny, Splitting properties in Archimedean ℓ-groups. *J. Austral. Math. Soc. 23 (ser A):* 247-256 (1977).

2. G. Birkhoff, *Theory of structures*. Moscow (1948).

3. B.Z. Vulih, *Introduction to the theory of partially ordered spaces*. Moscow (1961).

ANTI-f-RINGS

Robert Ross Wilson

California State University
Long Beach, California

1. INTRODUCTION

Among ℓ-rings the most widely studied have surely been the f-rings. Being subdirect products of totally ordered rings they may be viewed as separated strands (components) whose addition, multiplication and order are independent of each other. A well known consequence of this is the fact that multiplication by positive elements distributes over lattice operations. (True in total orders and preserved under subdirect product.) Since this is the only distributivity not true of all ℓ-rings, we will merely call it distributivity. For completeness and for reference we spell it out:

(*) if $a \geq 0$, then $a(b \wedge c) = ab \wedge ac$, $a(b \vee c) = ab \vee ac$, $(b \wedge c)a = ba \wedge ca$ and $(b \vee c)a = ba \vee ca$.

Naturally (*) holds whenever b and c are comparable, since multiplication by positive elements is a homomorphism on chains. We consider in this paper a class of ℓ-rings for which this restricted sense of distributivity is all that holds. That is, an *anti-f-ring* is an ℓ-ring in which (*) hold only when b and c are comparable.

Of course, the additive ℓ-group structure is still a subdirect product of totally ordered ℓ-groups, so that the 'strands' are still there as far as addition and order are concerned. However, in anti-f-rings, either they are totally ordered (so there is only one 'strand') or the multiplicative structure so intermingles the 'strands' that distributivity breaks down.

2. MAIN RESULT

Since most of my interest has been directed toward constructing lattice orders on fields and other rings by algebraic extension the main results concern such extensions. All rings considered in this section are commutative integral domains, but similar results hold for more general rings.

First we define an acceptable algebraic extension recursively.

(i) If $\langle R, P_R \rangle$ is an ℓ-ring and α is algebraic of order n over R, then $\langle R(\alpha), P_R(\alpha) \rangle$ is acceptable iff $\alpha^n \in P_R(\alpha)$.

(ii) If $\langle S, P_S \rangle$ is the union of a chain of acceptable algebraic extensions of $\langle R, P_R \rangle$, then $\langle S, P_S \rangle$ is an acceptable algebraic extension of $\langle R, P_R \rangle$.

Note that in case (i) $\langle R(\alpha), P_R(\alpha) \rangle$ is an ℓ-ring, and in case (ii) if $S = \bigcup_\gamma S_\gamma$ with $\langle S_\gamma, P_\gamma \rangle$ acceptable, then $P_S = \bigcup_\gamma P_\gamma$ makes $\langle S, P_S \rangle$ an ℓ-ring.

LEMMA 1. If $\langle R, P_R \rangle$ is an anti-f-ring and $R(\alpha)$ is an acceptable algebraic extension of R, then $\langle R(\alpha), P_R(\alpha) \rangle$ is an anti-f-ring.

The proof will be given below.

THEOREM 2. If $\langle R, P_R \rangle$ is an anti-f-ring and $\langle S, P_S \rangle$ is an acceptable algebraic extension of $\langle R, P_R \rangle$, then $\langle S, P_S \rangle$ is an anti-f-ring.

Proof. An obvious and easy induction on Lemma 1.

COROLLARY 3. Any formally real field admits an ℓ-order under which it is an anti-f-ring.

Proof. This follows from results in [19] and [20].

Proof of LEMMA 1. It is sufficient to consider the restricted sense of distributivity $(t \vee 0)s = ts \vee 0$ whenever $s \geq 0$.

First let $s = a_i\alpha^i$ for some fixed i, $0 \leq i \leq n - 1$. Then suppose $t = \Sigma_j b_j \alpha^j$, so that $t \vee 0 = \Sigma_j (b_j \vee 0)\alpha^j$. Now $ts = \Sigma_j b_j \alpha^j a_i \alpha^i = \Sigma_j b_j a_i \alpha^j \alpha^i$, so $ts \vee 0 = \Sigma_j (B_j a_i \vee 0)\alpha^{j+i}$ and $(t \vee 0)s = \Sigma_j (b_j \vee 0)a_i\alpha^{j+i}$ similarly. Thus $(t \vee 0)s = ts \vee 0$ iff $(b_j a_i \vee 0) = (b_j \vee 0)a_i$ for all j. Since a_i is an arbitrary positive element of R, it follows that each b_j is comparable with 0.

Suppose now that $b_j \geq 0$ and $b_k \leq 0$, then let $s = b_j\alpha^j - b_k\alpha^k$. Now the coefficients of α^{j+k} in the two sides of the distributivity equation are $(b_k \vee 0)b_j - (b_j \vee 0)b_k = - b_jb_k$ and $(b_jb_k - b_kb_j) \vee 0 = 0$, so equality forces one of the b's to be 0. Thus t must be comparable with 0.

Though the concept of anti-f-ring applies nicely to the process of ℓ-order construction by algebraic extension it is unfortunately, and not surprisingly, not useful when applying other standard techniques of ring construction. For example, the direct product of anti-f-rings is never an anti-f-ring and the homomorphic image of an anti-f-ring is not, in general, an anti-f-ring.

3. FURTHER EXAMPLES

EXAMPLE 4. Consider formal power series rings $W = W(\Gamma,R)$ constructed as in [17], where Γ is a root system and a strictly po-semigroup which is commutative and cancellative and where R is the real field. A proof closely modeled on that of Lemma 1 shows that such rings are anti-f-rings.

EXAMPLE 5. The dual numbers (example 9e of [1]) are an anti-f-ring. Many other algebras which have a similar crossing over of the strands in their multiplication generalize this example.

These examples along with Corollary 3 seem to indicate that anti-f-rings are at least important in the study of ℓ-fields and may have much broader interest.

REFERENCES

BASIC PAPER

1. Garrett Birkhoff and Richard S. Pierce, Lattice-ordered rings. *An. Acad. Brasil Ci. 28:* 41-69 (1956).

FUNCTION RINGS

2. S. J. Bernau, Unique representations of Archimedean lattice groups and normal Archimedean lattice rings. *Proc. London Math. Soc. 15:* 599-631 (1965).

3. Paul Conrad, The hulls of representable ℓ-groups and f-rings. *J. Austral. Math. Soc. 16:* 385-415 (1973).

4. John Dauns, Representation of f-rings. *Bull. Amer. Math. Soc. 74:* 249-252 (1968).

5. John Dauns, Representation of ℓ-groups and f-rings. *Pacific J. Math. 34:* 365-369 (1970).

6. Anthony W. Hager, Some remarks on the tensor product of function rings. *Math. Z. 92:* 210-224 (1966).

7. Melvin Henriksen and John R. Isbell, Lattice-ordered rings and function rings. *Pacific J. Math. 12:* 533-565 (1962).

8. Donald G. Johnson, A structure theory for a class of lattice-ordered rings. *Acta Math. 104:* 163-215 (1960).

9. Donald G. Johnson, On a representation theory for a class of Archimedean lattice-ordered algebras. *Proc. London Math. Soc. 12:* 207-225 (1962).

10. Donald G. Johnson, The completion of an Archimedean f-ring. *J. London Math. Soc. 40:* 493-496 (1965).

11. Joseph E. Kist, Representation of Archimedean function rings. *Illinois J. Math. 7:* 269-278 (1963).

12. John E. Mack and Donald G. Johnson, The Dedekind completion of C(X). *Pacific J. Math. 20:* 231-243 (1967).

13. Phillip Nanzetta, Maximal lattice-ordered algebras of continuous functions. *Fund. Math. 63:* 53-75 (1968).

14. Phillip Nanzetta, On the lattice D(X). *Illinois J. Math. 13:* 145-154 (1969).

15. Richard S. Pierce, Radicals in function rings. *Duke Math J. 23:* 253-261 (1956).

16. Stuart A. Steinberg, Finitely-valued f-modules. *Pacific J. Math. 40:* 723-737 (1972).

LATTICE-ORDERED FIELDS

17. Paul Conrad and John Dauns, An embedding theorem for lattice-ordered fields. *Pacific J. Math. 30:* 385-398 (1969).

18. Robert Ross Wilson, Lattice orders on real fields. Doctoral Dissertation, University of California, Los Angeles, 1974.

19. Robert Ross Wilson, Lattice ordering on the real field. *Pacific J. Math. 63:* 571-577 (1976).

20. Robert Ross Wilson, Lattice orders on fields and certain rings. *Symposia Mathematica. 21:* 358-364 (1977).

MISCELLANEOUS ORDERED RINGS

21. Ralph DeMarr, Partially ordered fields. *Amer. Math. Monthly. 74:* 418-420 (1967).

22. Ralph DeMarr, On partially ordering operator algebras. *Canad. J. Math. 20:* 636-643 (1967).

23. Ralph DeMarr, A generalization of the Perron-Frobenius Theorem. *Duke Math. J. 37:* 113-120 (1970).

24. Laszlo Fuchs, *Partially ordered algebraic systems.* Pergamon Press, 1963.

25. Anthony W. Hagar, Phillip Nanzetta and Donald Plank, Inversion in a class of lattice-ordered algebras. *Colloq. Math. 24:* 225-234 (1971/72).

26. Melvin Henriksen and John R. Isbell, Residue class fields of lattice-ordered algebras. *Fund. Math. 50:* 107-117 (1961/62).

27. Melvin Henriksen and Donald G. Johnson, On the structure of a class of Archimedean lattice-ordered rings. *Fund. Math. 50:* 73-94 (1961/62).

28. John R. Isbell, A structure space for certain lattice-ordered groups and rings. *J. London Math. Soc. 40:* 63-71 (1965).

A CLASS OF PARTIALLY ORDERED SETS WHOSE GROUPS OF ORDER AUTOMORPHISMS ARE LATTICE-ORDERED *

Maureen A. Bardwell

St. Norbert College
DePere, Wisconsin

In Birkhoff's book entitled *Lattice Theory*, the following question was posed: Classify those lattice-ordered groups (ℓ-groups) which are isomorphic (both as groups and as lattices) to the ℓ-group of order automorphisms of a totally ordered set (chain). The ℓ-group of order automorphisms of a chain has been carefully studied in recent years, primarily because every ℓ-group can be ℓ-embedded in such a group [3]. As a result of the research done, the ℓ-group of order automorphisms of a chain has been shown to possess a number of characteristics which an arbitrary ℓ-group may not have, and Birkhoff's original question has been answered.

Now let Ω denote an arbitrary partially ordered set (po-set) and let $A(\Omega)$ denote the group of order automorphisms of Ω. Then, $A(\Omega)$ is a partially ordered group (po-group) if we order $A(\Omega)$ by: for f,g in $A(\Omega)$, $f \leq g$ if and only if for all $\alpha \in \Omega$, $\alpha f \leq \alpha g$. In many cases, the partial order that Ω induces on $A(\Omega)$ is a lattice order, and so $A(\Omega)$ is an ℓ-group. A generalization of Birkhoff's original question is: Classify the po-sets Ω and the groups $A(\Omega)$ such that $A(\Omega)$ is an ℓ-group under the induced order.

Because of the generality of this problem, prior to the research connected with this work, no real progress had been made towards a solution. We will study this problem by placing a limiting hypothesis on the ℓ-groups we will consider. When Ω is totally ordered, the ℓ-group $A(\Omega)$ is laterally complete and completely distributive [2]. The group also has the property

---

* This paper grew out of the author's doctoral dissertation which was directed by Professor W. Charles Holland of Bowling Green State University. The author wishes to thank Professor Holland, without whose assistance this article would not have been possible.

that positive elements of $A(\Omega)$ are algebraically disjoint if and only if they have disjoint supports as permutations [3]. This property is called the *disjoint support property*.

In [1], the present author gives a complete classification of all the po-sets Ω such that (1) $A(\Omega)$ is an ℓ-group with the disjoint support property. One of the basic results of that paper is that condition (1) is equivalent to (2) every orbit of $A(\Omega)$ in Ω is a chain. In the present work, we derive another condition which is equivalent to the previous two and which deals with the prime subgroups of $A(\Omega)$. We show that the ℓ-groups we are considering are laterally complete, and we illustrate the basic classification procedure by several detailed examples. One of these is an example of a po-set Ω such that $A(\Omega)$ has an infinite number of orbits in Ω, and $A(\Omega)$ is uniquely transitive on each of its orbits. We also consider the question of closed stabilizers and complete distributivity in these groups.

If Ω is any po-set, $A(\Omega)$ denotes the collection of all permutations f of Ω such that both f and f^{-1} preserve the order relation on Ω. For any f in $A(\Omega)$, define the *support* of f, denoted by supp f, to be $\{\alpha \in \Omega | \alpha f \neq \alpha\}$. For elements f and g of any ℓ-group G, f and g are said to be *algebraically disjoint* if and only if $f \wedge g = e$ (where e denotes the group identity). An ℓ-group G is said to be *laterally complete* if and only if any collection of algebraically disjoint positive elements has a supremum. G is *completely distributive* if for $g_{ij} \in G$, $\bigwedge_I \bigvee_J g_{ij} = \bigvee_{f \in J^I} \bigwedge_I g_{if(i)}$, provided that the indicated sups and infs exist. If G is an ℓ-group, H is an ℓ-subgroup of G if H is both a subgroup and a sublattice of G. H is *convex* in G if $g \in G$, $h \in H$, and $e \leq g \leq h$ imply that $g \in H$. H is a *prime* subgroup of G if H is a convex ℓ-subgroup of G and if $f \wedge g \in H$, with $f, g \in G$, implies $f \in H$ or $g \in H$. If Ω is a po-set, $A(\Omega)$ is its group of order automorphisms, and G is a subgroup of $A(\Omega)$, for $\alpha \in \Omega$ define the *stabilizer subgroup* G_α to be $\{g \in G | \alpha g = \alpha\}$. If G is any ℓ-group and H is an ℓ-subgroup of G, H is *closed* if $g = \sup\{h_i | i \in I\}$ with g in G and h_i in H for each i in I implies that $g \in H$.

We now present a theorem which gives several equivalences to the statement that Ω is a po-set such that $A(\Omega)$ is an ℓ-group with disjoint support property. As stated in [1] the reasons for choosing the disjoint support property as a limiting hypothesis are that this is an important property of the ℓ-group of order automorphisms of a chain, and that in these groups the algebraic notion of disjointness and the geometric notion of disjointness coincides.

THEOREM 1. Suppose Ω is a partially ordered set and $A(\Omega)$ is the group of order automorphisms of Ω. Then the following are equivalent:

(1) $A(\Omega)$ is an ℓ-group having the disjoint support property.

(2) Every orbit of $A(\Omega)$ in Ω is a chain.

(3) $A(\Omega)$ is an ℓ-group and for all $\alpha \in \Omega$ and $f,g \in A(\Omega)$, $\alpha(f \vee g) = \max\{\alpha f, \alpha g\}$ and $\alpha(f \wedge g) = \min\{\alpha f, \alpha g\}$.

(4) $A(\Omega)$ is an ℓ-group and for all $\alpha \in \Omega$, G_α is a prime subgroup of $A(\Omega)$.

Proof. (1) $\leftrightarrow$ (2) $\leftrightarrow$ (3) is proved in [1]. To prove (3) $\rightarrow$ (4), we may use the same techniques used to prove this statement as when Ω is a chain. Now we check (4) $\rightarrow$ (3). The only thing that needs to be done is to check how we compute sups and infs. Fix α in Ω and f,g in $A(\Omega)$. For ease of notation, set $f \wedge g = k$. Since $\alpha(f \wedge g) = \alpha k$, $\alpha(f \wedge g)k^{-1} = \alpha$, i.e., $\alpha(fk^{-1} \wedge gk^{-1}) = \alpha$. $\alpha(fk^{-1} \wedge gk^{-1}) \in G_\alpha$ and G_α is prime so $fk^{-1} \in G_\alpha$ or $gk^{-1} \in G_\alpha$. If $\alpha fk^{-1} = \alpha$ then $\alpha f = \alpha(f \wedge g)$. If $\alpha gk^{-1} = \alpha$ then $\alpha g = \alpha(f \wedge g)$. Thus, either $\alpha f \leq \alpha g$ and $\alpha(f \wedge g) = \alpha f = \min\{\alpha f, \alpha g\}$ or $\alpha g \leq \alpha f$ and $\alpha(f \wedge g) = \alpha g = \min\{\alpha f, \alpha g\}$. The previous argument shows that αf and αg are comparable, so $\alpha(f \vee g) = \max\{\alpha f, \alpha g\}$.

In the general theory of ℓ-groups, the prime subgroups play a significant role. When studying permutation groups, one is naturally led to consider the stabilizer subgroups. Condition (4) therefore again assures us that the disjoint support property is a useful hypothesis. To give an example of a po-set Ω such that $A(\Omega)$ is an ℓ-group but $A(\Omega)$ does not have prime stabilizers, consider $\mathbb{R} \boxplus \mathbb{Z}$, the cardinal product of $\mathbb{R}$ with $\mathbb{Z}$. Let $+n$ denote translation by the real number n. Then $A(\mathbb{R} \boxplus \mathbb{Z}) = A(\mathbb{R}) \times A(\mathbb{Z})$, $(+1,+0)$ and $(+0,+1)$ are not in any stabilizer subgroup, but $(+1,+0) \wedge (+0,+1) = (+0,+0)$, which is in every stabilizer subgroup.

In [1], we showed that condition (3) implies that if $\{0_i | i \in I\}$ is the collection of orbits of $A(\Omega)$ in Ω, $A(\Omega)$ can be considered as an ℓ-subgroup of the cardinal product of the $\{A(0_i) | i \in I\}$, where the action of $A(\Omega)$ on Ω is the natural one. Obviously, condition (2) implies that if $A(\Omega)$ is transitive on Ω, Ω is a homogeneous chain. Our original hypotheses also imply the following.

PROPOSITION 2. Suppose Ω is a po-set such that $A(\Omega)$ is an ℓ-group having the disjoint support property. Then $A(\Omega)$ is laterally complete.

Proof. We may use a similar proof to the one given when Ω is a chain.

The method of classification proceeded as follows. We assumed $A(\Omega)$ had precisely two orbits in Ω, O_1 and O_2. Let P_1 denote the projection of elements of $A(\Omega)$ onto their first coordinates and let P_2 denote the projection of elements of $A(\Omega)$ onto their second coordinates. We define convex P_1-congruences $\mathcal{C}_{1,2}$ and $\mathcal{C}^{1,2}$ on O_1 by: for $s_1, s_2, \in O_1$, $s_1\mathcal{C}_{1,2}s_2$ if and only if s_1 and s_2 have the same collection of lower bounds in O_2, and $s_1\mathcal{C}^{1,2}s_2$ if and only if s_1 and s_2 have the same collection of upper bounds in O_2. Let $\mathcal{U}_1$ denote the universal congruence on O_1 and $\mathcal{E}_1$ denote the trivial congruence on O_1. Define $\mathcal{E}_2$, $\mathcal{U}_2$, $\mathcal{C}_{2,1}$, and $\mathcal{C}^{2,1}$ on O_2 similarly. Theorem 1 completely describes the order relations between points in the same orbit and these congruences completely describe the possible order relations which may exist between points of different orbits. Since P_1 is transitive on O_1, the convex P_1-congruences form a tower under inclusion [2]. Similarly for the convex P_2-congruences on O_2. Thus, to perform the classification, we had to examine the possible containment relations which could exist between the congruences on O_1 and between the congruences on O_2. By making appropriate reductions, this was reduced to examining six basic types of po-sets. For a listing of these types, the reader is referred to [1].

The real motivation for the classification theorems given for these types is the examples. In this paper, we offer no abstract classification, but will now present in detail several specific examples. We make one notational convention. We will consider examples where one or more of $\mathcal{C}^{1,2}$ and $\mathcal{C}_{1,2}$ (and/or $\mathcal{C}^{2,1}$ or $\mathcal{C}_{2,1}$) will be proper convex congruences. Suppose, for instance, that $\mathcal{E}_1 \subset \mathcal{C}_{1,2} \subset \mathcal{U}_1$. Then, if $s\mathcal{C}_{1,2}$ denotes a typical $\mathcal{C}_{1,2}$-class, the ℓ-permutation group (P_1, O_1) may be ℓ-embedded into $(A(s\mathcal{C}_{1,2}), s\mathcal{C}_{1,2})\mathrm{Wr}$ $(A(O_1/\mathcal{C}_{1,2}), O_1/\mathcal{C}_{1,2})$ [2]. We will most often be concerned with the "group part" of these ℓ-embeddings and, since the appropriate sets will be clear from context, we will write $P_1 \subseteq A(s\mathcal{C}_{1,2})\mathrm{Wr}A(O_1/\mathcal{C}_{1,2})$. For notation and general terminology on ordered wreath products, the reader is referred to [2].

EXAMPLE 3. ($\mathcal{C}_{1,2} = \mathcal{C}^{1,2} = \mathcal{U}_1$ on O_1 and $\mathcal{C}_{2,1} = \mathcal{C}^{2,1} = \mathcal{U}_2$ on O_2)

Let $O_1 = Q$, $O_2 = \Pi$, the irrational numbers, and let $\Omega = O_1 \cup O_2$. Define an order relation $\leq$ on Ω by $\leq|_{O_1} = \leq_Q$, $\leq|_{O_2} = \leq_{\Pi}$, and every point of O_1 is unrelated to every point of O_2 under $\leq$. Pictorally, Ω appears as follows:

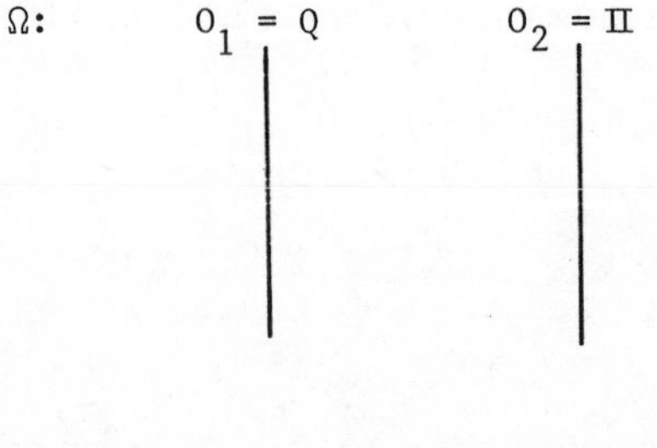

We claim the orbits of $A(\Omega)$ in Ω are O_1 and O_2. By way of contradiction, suppose there exists $f \in A(\Omega)$ and $q \in O_1$ such that $qf \in O_2$. Since f is order preserving, f must map unrelated points to unrelated points and related points to related points. Thus $qf \in O_2$ implies $O_1 f = O_2$ and $O_2 f = O_1$. But f cannot switch the two chains since O_1 is not order-isomorphic to O_2. Thus no such f exists, and it follows that $A(\Omega) \subseteq A(Q) \times A(\mathbb{I})$. An easy computation shows each element of $A(Q) \times A(\mathbb{I})$ is order preserving on Ω, so $A(\Omega) = A(Q) \times A(\mathbb{I})$.

Before proceeding further, we make some notational conventions. For any chain O, $\overline{O}$ will denote its completion by Dedekind cuts. For any $f \in A(O)$, f can be uniquely extended to a permutation $\overline{f} \in A(\overline{O})$ [4]. If $s \in O_1$, let $L_{1,2}(s)$ denote $\{t \in O_2 | t \leq_\Omega s\}$ and $U_{1,2}(s)$ denote $\{t' \in O_2 | s \leq_\Omega t'\}$. For $t \in O_2$, define $L_{2,1}(t)$ and $U_{2,1}(t)$ similarly.

EXAMPLE 4. $(\mathcal{E}_1 = \mathcal{C}_{1,2} \subset \mathcal{C}^{1,2} = \mathcal{U}_1$ on O_1 and $\mathcal{E}_2 = \mathcal{C}^{2,1} \subset \mathcal{C}_{2,1} = \mathcal{U}_2$ on $O_2)$

Let $O_1 = Q$ and $O_2 = \mathbb{I}$. Let $\mathcal{X}_{1,2} : O_1 \to \overline{O}_2$ denote the identity map, and $\mathcal{X}^{2,1} : O_2 \to \overline{O}_1$ denote the identity map. Partially order the set $\Omega = O_1 \dot{\cup} O_2$ as follows: $\leq_{1_{O_1}} = \leq_Q$, $\leq_{1_{O_2}} = \leq_{\mathbb{I}}$, for each $s \in O_1$, $L_{1,2}(S) = \{t \in O_2 | t \leq s\, \mathcal{X}_{1,2}\}$ and $U_{1,2}(s) = \emptyset$, and for each $t \in O_2$, let $L_{2,1} = \emptyset$ and $U_{2,1}(t) = \{s \in O_1 | t\, \mathcal{X}^{2,1} \leq s\}$. Pictorally, Ω appears as follows:

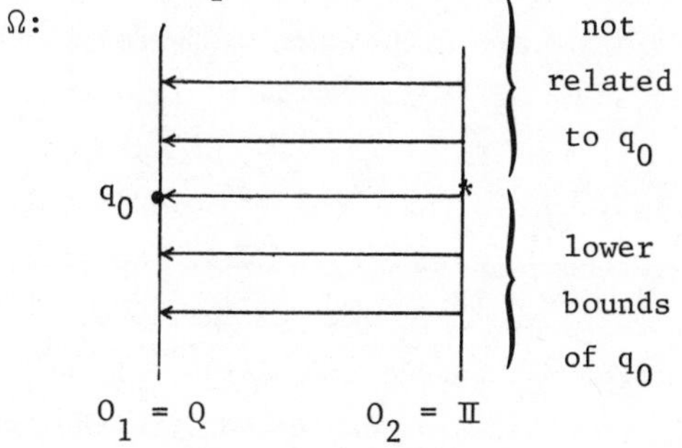

Note that elements of $A(\Omega)$ must map points of O_1 to points of O_1 and points of O_2 to points of O_2 since points of O_1 have two chains underneath them whereas points of O_2 only have one chain underneath them. It now follows that $A(\Omega) \subseteq A(Q) \times A(\mathbb{I})$.

To finish computing $A(\Omega)$, notice $\mathcal{X}_{1,2}$ can be extended to an order-isomorphism $\overline{\mathcal{X}}_{1,2}$ (the identity map) from $\overline{O}_1 = \mathbb{R}$ to $\overline{O}_2 = \mathbb{R}$. We may define an ℓ-isomorphism $\overline{*}$: $A(\overline{O}_1) \to A(\overline{O}_2)$ by: for $f \in A(\overline{O}_1)$, $f^{\overline{*}} = \overline{\mathcal{X}}_{1,2}^{-1} f\, \overline{\mathcal{X}}_{1,2}$

Set $G_{1,2} = \{f \in A(Q) | \mathbb{I}f = \mathbb{I}\} = A(Q)$ and $G_{2,1} = \{g \in A(\mathbb{I}) | Qg = Q\} = A(\mathbb{I})$. By composing extension, $\overline{*}$, and restriction, we define an ℓ-isomorphism $*$: $A(Q) \to A(\mathbb{I})$. It is then straightforward to check that $A(\Omega) = \{(f, f^*) | f \in A(Q)\}$. Since $A(\Omega)$ is transitive on O_1 and O_2, the orbits of $A(\Omega)$ in Ω are the chains O_1 and O_2. Thus $A(\Omega)$ is an ℓ-group with the disjoint support property. The desired congruence structure follows from the properties of $\mathcal{X}_{1,2}$ and $\mathcal{X}^{2,1}$.

EXAMPLE 5. ($\mathcal{E}_1 = \mathcal{C}_{1,2} \subset \mathcal{C}^{1,2} = \mathcal{U}_1$ on O_1 and $\mathcal{E}_2 \subset \mathcal{C}^{2,1} \subset \mathcal{C}_{2,1} = \mathcal{U}_2$ on O_2)

Let $O_1 = \mathbb{Z}$ and $O_2 = \mathbb{R} \overleftarrow{\times} \mathbb{Z}$. Define $\mathcal{X}_{1,2}: O_1 \to \overline{O}_2$ by: for n_0 in $\mathbb{Z}$, $n_0\mathcal{X}_{1,2} = \sup\{(r,n_0) \in \mathbb{R} \overleftarrow{\times} \mathbb{Z} \mid r \in \mathbb{R}\}$ and $\mathcal{X}^{2,1}: O_2 \to \overline{O}_1$ by: for (r,n_0) in $\mathbb{R} \overleftarrow{\times} \mathbb{Z}$, $(r,n_0)\mathcal{X}^{2,1} = n_0$. $\mathcal{X}_{1,2}$ is one-to-one and order preserving, but $O_1\mathcal{X}_{1,2}$ is not dense in $\overline{O}_2$. Moreover, $O_1\mathcal{X}_{1,2}$ is a subset of an orbit of $A(O_2)$ in $\overline{O}_2$. $\mathcal{X}^{2,1}$ is not one-to-one, but is order preserving, and $O_2\mathcal{X}^{2,1}$ is dense in $\overline{O}_1 = O_1$.

Define a partial order on the set $\Omega = O_1 \dot{\cup} O_2$ by: $\leq_{1_{O_1}} = \leq_{\mathbb{Z}}$, $\leq_{1_{O_2}} = \leq_{\mathbb{R} \overleftarrow{\times} \mathbb{Z}}$, for each n_0 in O_1 set $L_{1,2}(n_0) = \{(r,n) \in O_2 \mid (r,n) \leq_{\overline{O}_2} n_0\mathcal{X}_{1,2}\}$ and $U_{1,2} = \phi$ and for each (r,n_0) in O_2, set $L_{2,1}((r,n_0)) = \phi$ and $U_{2,1}((r,n_0)) = \{n \in O_2 \mid n_0 \leq_{\mathbb{Z}} n\}$.

We may use the same argument as was used in Example 4 to show elements of $A(\Omega)$ must map points of O_1 to points of O_1 and points of O_2 to points of O_2. Thus $A(\Omega) \subseteq A(O_1) \times A(O_2)$. Let $\operatorname{Ker}\mathcal{X}^{2,1}$ denote the convex P_2-convergence defined by $(r_1,n_1)\mathcal{X}^{2,1} = n_1 = n_2 = (r_2,n_2)\mathcal{X}^{2,1}$. $O_2/\operatorname{Ker}\mathcal{X}^{2,1}$ is order-isomorphic to O_1 and $A(O_2) = A((r,n)\operatorname{Ker}\mathcal{X}^{2,1}) \operatorname{Wr} A(O_2/\operatorname{Ker}\mathcal{X}^{2,1}) = A(\mathbb{Z}) \operatorname{Wr} A(\mathbb{R})$. Let * denote the identity ℓ-isomorphism from $A(O_1)$ to $A(O_2/\operatorname{Ker}\mathcal{X}^{2,1})$. Then, a straightforward computation shows $A(\Omega) = \{(f,(\phi:f^*)) \in A(\mathbb{Z}) \times [A(\mathbb{R}) \operatorname{Wr} A(\mathbb{Z})]\}$, and it follows that the orbits of $A(\Omega)$ in Ω are O_1 and O_2. $A(\Omega)$ is thus an ℓ-group having the disjoint support property, and the congruence structure follows from the properties of the map.

EXAMPLE 6. ($\mathcal{E}_1 = \mathcal{C}_{1,2} = \mathcal{C}^{1,2} \subset \mathcal{U}_1$ and $\mathcal{E}_2 = \mathcal{C}_{2,1} = \mathcal{C}^{2,1} \subset \mathcal{U}_2$)

Let $O_1 = Q$ and $O_2 = \mathbb{I}$. Define $\mathcal{X}_{1,2}: O_1 \to \overline{O}_2$ and $\mathcal{X}^{1,2}: O_1 \to \overline{O}_2$ via: for q_0 in O_1, $q_0\mathcal{X}_{1,2} = q_0$ and $q_0\mathcal{X}^{1,2} = q_0 + 1$. Then $\mathcal{X}_{1,2}$ and $\mathcal{X}^{1,2}$ are one-to-one and order preserving, and $O_1\mathcal{X}_{1,2}$ and $O_1\mathcal{X}^{1,2}$ are dense subsets of $\overline{O}_2$. Define $\mathcal{X}_{2,1}: O_2 \to \overline{O}_1$ and $\mathcal{X}^{2,1}: O_2 \to \overline{O}_1$ by: for i_0 in O_2, $i_0\mathcal{X}_{2,1} = i_0 - 1$ and $i_0\mathcal{X}^{2,1} = i_0$. The properties of the maps are analagous to those of $\mathcal{X}_{1,2}$ and $\mathcal{X}^{1,2}$. Partially order the set $\Omega = O_1 \dot{\cup} O_2$ as follows: $\leq_{1_{O_1}} = \leq_Q$, $\leq_{1_{O_2}} = \leq_{\mathbb{I}}$, for each q_0 in Q set $L_{1,2}(q_0) = \{i \in O_2 \mid i \leq_{\mathbb{R}} q_0\mathcal{X}_{1,2}\}$ and set $U_{1,2}(q_0) = \{i \in O_2 \mid q_0\mathcal{X}^{1,2} \leq_{\mathbb{R}} i\}$, and for each i_0 in O_2 set $L_{2,1}(i_0) =$

$\{q \in O_1 \mid q \leq i_0\mathcal{X}_{2,1}\}$ and set $U_{2,1}(i_0) = \{q \in O_1 \mid i_0\mathcal{X}^{2,1} \leq_{\mathbb{R}} q\}$. Pictorally, Ω is as follows:

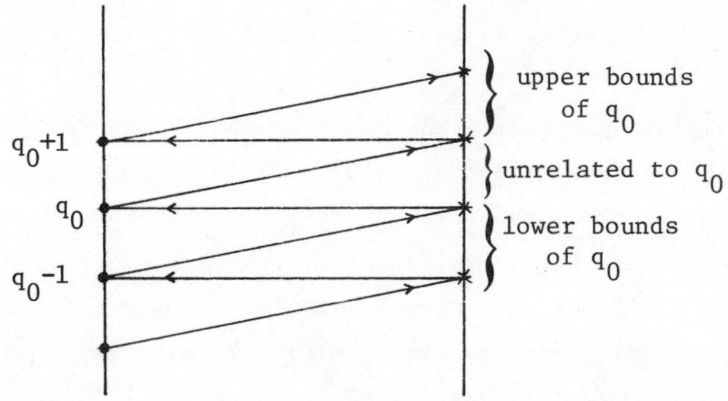

We claim that elements of $A(\Omega)$ must map points of O_1 to points of O_1 and points of O_2 to points of O_2. By noting that points that are unrelated must remain unrelated under the action of a function in $A(\Omega)$ and by noting that we can get to the ends of O_1 and O_2 by "weaving back and forth" by the maps $\mathcal{X}^{1,2}$ and $\mathcal{X}^{2,1}$, we can show that if we map any point of O_1 to a point of O_2, we must map all points of O_1 to O_2 and all points of O_2 to O_1. This cannot be done, however, since O_1 is not order-isomorphic to O_2. Hence, the claim is established. We may therefore consider $A(\Omega)$ as a subgroup of $A(O_1) \times A(O_2)$. We show it is an ℓ-subgroup. $\mathcal{X}_{1,2}$ may be extended to an order-isomorphism $\overline{\mathcal{X}}_{1,2}$ (the identity map) from $\overline{O}_1 = \mathbb{R}$ to $\overline{O}_2 = \mathbb{R}$. $\overline{\mathcal{X}}_{1,2}$ induces the identity ℓ-isomorphism $\overline{*}$: $A(\overline{O}_1) \to A(\overline{O}_2)$ by: for $\overline{f}$ in $A(\overline{O}_1)$, $\overline{f*} = \overline{\mathcal{X}}_{1,2}^{-1}\,\overline{f}\,\overline{\mathcal{X}}_{1,2}$. Note that $\overline{\mathcal{X}}^{2,1}\overline{\mathcal{X}}^{1,2} = +1$, the permutation which is translation by 1, and $\overline{\mathcal{X}}^{1,2}\overline{\mathcal{X}}^{2,1} = +1$. Set $G_{1,2} = \{f \in A(Q) \mid \mathbb{I}\overline{f} = \mathbb{I}$ and $\overline{f}(+1) = (+1)\overline{f}\} = \{f \in A(Q) \mid f(+1) = (+1)f\}$, and set $G_{2,1} = \{g \in A(\mathbb{I}) \mid g(+1) = (+1)g\}$. By composing extension, $\overline{*}$, and restriction we may define an ℓ-isomorphism $*$: $G_{1,2} \to G_{2,1}$. Set $G = \{(f,f^*) \mid f \in G_{1,2}\}$. Clearly $A(\Omega) \subseteq G$. A straightforward computation shows each element of G is order preserving on Ω. Thus $G = A(\Omega)$. G is transitive on O_1 and O_2. Hence every orbit of $A(\Omega)$ in Ω is a chain, and it follows that $A(\Omega)$ is an ℓ-group with disjoint support property. The congruence structure is clearly what was claimed.

EXAMPLE 7. ($\mathcal{E}_1 = \mathcal{C}_{1,2} \subset \mathcal{C}^{1,2} \subset \mathcal{U}_1$ on O_1 and $\mathcal{E}_2 = \mathcal{C}_{2,1} \subset \mathcal{C}^{2,1} \subset \mathcal{U}_2$ on O_2)

Let $\Gamma = \{0,1,2,\ldots,\infty\}$ and $\Delta = \{0,1,2,\ldots,\infty\}$ be two index sets. For each $\gamma \in \Gamma$, we define chains S_γ as follows: if $\gamma \equiv 0 \pmod 2$ let $S_\gamma = Q$; if $\gamma \equiv 1 \pmod 2$ let $S_\gamma = \mathbb{I}$; if $\gamma = \infty$ let $S_\gamma = \mathbb{R}$. Similarly, for each $\delta \in \Delta$, we define chains T_δ as follows: if $\delta \equiv 0 \pmod 2$ let $T_\delta = \mathbb{I}$; if $\delta \equiv 1 \pmod 2$ let $T_\delta = Q$; if $\delta = \infty$ let $T_\delta = \mathbb{R}$. Set $S' = \Pi\{S_\gamma \mid \gamma \in \Gamma\}$ and $T' = \Pi\{T_\delta \mid \delta \in \Delta\}$.

$\tilde{0}_S$ will denote the vector in S' which has the zero element of Q appearing in each coordinate whose index is congruent to 0 (modulo 2), $\sqrt{2}$ appearing in each coordinate whose index is congruent to 1 (modulo 2), and the zero element of $\mathbb{R}$ appearing in the ∞-coordinate. Similarly, $\tilde{0}_T$ will denote the vector in T' which has the $\sqrt{2}$ appearing in each coordinate whose index is congruent to 0 (modulo 2), the zero element of Q appearing in each coordinate whose index is congruent to 1 (modulo 2), and the zero element of $\mathbb{R}$ appearing in the ∞-coordinate. Let $\tilde{s} = (q_o, i_1, q_2, i_3, \ldots, r_\infty)$ denote a typical element of S' and $\tilde{t} = (i_o, q_1, i_2, q_3, \ldots r_\infty)$ denote a typical element of T'. Set $S = \{\tilde{s} \in S' | s_\gamma \neq 0_{S_\gamma}$ for only finitely many $\gamma \in \Gamma\}$ and set $T = \{\tilde{t} \in T' | t_\delta \neq 0_{T_\delta}$ for only finitely many $\delta \in \Delta\}$. By ordering Γ and Δ in the obvious way, we may define total orders of the sets S and T by ordering lexicographically from the right. S and T are homogeneous chains, A(S) is ℓ-isomorphic to $\mathrm{Wr}\{A(S_\gamma) | \gamma \in \Gamma\}$, and A(T) is ℓ-isomorphic to $\mathrm{Wr}\{A(T_\delta) | \delta \in \Delta\}$.

To construct the example let $O_1 = S$, $O_2 = T$, and $\Omega = O_1 \dot{\cup} O_2$. We define maps $\mathcal{X}^{1,2}$ and $\mathcal{X}_{1,2}$ from O_1 to $\overline{O}_2$ and maps $\mathcal{X}^{2,1}$ and $\mathcal{X}_{2,1}$ from O_2 to $\overline{O}_1$ in the following way. For $(q_0, i_1, q_2, i_3, \ldots, r_\infty)$ in O_1, let $(q_0, i_1, q_2, i_3, \ldots, r_\infty)\mathcal{X}^{1,2} = (i_1, q_2, i_3, \ldots, r_\infty + 1)$ in O_2 and $(q_0, i_1, q_2, i_3, \ldots, r_\infty)\mathcal{X}_{1,2} = \sup\{(i, q_0, i_1, q_2, \ldots, r_\infty) \in O_2 | i \in T_0\}$. For $(i_0, q_1, i_2, q_3, \ldots, r_\infty)$ in O_2, let $(i_0, q_1, i_2, q_3, \ldots, r_\infty)\mathcal{X}^{2,1} = (q_1, i_2, q_3, \ldots, r_\infty)$ in O_1 and $(i_0, q_1, i_2, q_3, \ldots, r_\infty)\mathcal{X}_{2,1} = \sup\{(q, i_0, q_1, i_2, \ldots, r_\infty - 1) \in O_1 | q \in S_0\}$. By using these maps, define a partial order of the set Ω as was done in Example 6. Then, by using the same argument as was given in Example 6, we show that elements of $A(\Omega)$ must map points of O_1 to points of O_1 and points of O_2 to points of O_2. Thus, we may consider $A(\Omega)$ as a subgroup of $A(O_1) \times A(O_2)$. We proceed to show it is an ℓ-subgroup of this ℓ-group, and that the po-set Ω has the properties desired.

To compute the group $A(\Omega)$, we must specify which members of $A(O_1) \times A(O_2)$ preserve order on Ω. There is a countable tower of natural proper convex equivalence relations $\mathcal{S}_1 \subseteq \mathcal{S}_2 \subseteq \cdots \subseteq \mathcal{S}_\infty$ on O_1. $\mathcal{S}_\infty$ is the join of the convex equivalence relations $\mathcal{S}_1, \mathcal{S}_2, \mathcal{S}_3, \ldots$. Similarly, there is a countable tower of natural proper convex equivalence relations $\mathcal{T}_1 \subseteq \mathcal{T}_2 \subseteq \cdots \subseteq \mathcal{T}_\infty$ on O_2. These relations give rise to the components used to construct the generalized wreath products mentioned previously.

For the remainder of the proof it will simplify the notation to identify denoted by (f,g). Choose $\tilde{s} = (q_0, i_1, q_2, i_3, \ldots r^\infty)$ in O_1. We recall that $\tilde{s}\mathcal{X}^{1,2} = (i_1, q_2, i_3, \ldots, r_\infty + 1)$ and $\tilde{s}\mathcal{X}_{1,2} = \sup\{(i, q_0, i_1, q_2, i_3, \ldots r_\infty) | i \in T_0\}$. Note that $\tilde{s}\mathcal{X}_{1,2}$ is the supremum of a $\mathcal{T}_1$-class. We fix $\tilde{s}\mathcal{S}_1$. For $\tilde{s}'$ in $\tilde{s}\mathcal{S}_1$ and each γ in Γ, $\gamma \geq 1$, $s_\gamma' = s_\gamma$. For $\tilde{s}'$ in $\tilde{s}\mathcal{S}_1$ and $\gamma = 0$, s_γ' may be any

element of $S_0 \cong Q$. It now follows that each $\tilde{s}'$ in $\tilde{s}\mathcal{S}_1$ has the same image under $\mathscr{X}^{1,2}$. Consider $(\tilde{s}\mathcal{S}_1)\mathcal{X}_{1,2} = \{\tilde{s}'\mathcal{X}_{1,2} | \tilde{s}' \in \tilde{s}\mathcal{S}_1\}$. This set consists of the supremum of all $\mathcal{T}_1$-classes which contain elements of O_2 of the form $(-,-,i_1,q_2,i_3,\ldots,r_\infty)$ (the two -'s can be replaced by any element of T_0 and any element of T_1). The action of a function on the supremum of a $\mathcal{T}_1$-class corresponds to an action of the class. By taking the union of the $\mathcal{T}_1$-classes mentioned, we obtain a convex $\mathcal{T}_2$-class. Since elements (f,g) of $A(\Omega)$ must respect $\mathscr{X}^{1,2}$ and $\mathcal{X}_{1,2}$, we see that it must be the case that $f_{1,(-,i_1,q_2,i_3,\ldots,r_\infty)} = g_{0,(i_1,q_2,i_3,\ldots,r_\infty+1)} = g_{2,(-,-,i_1,q_2,i_3,\ldots,r_\infty)}$. The point $(i_1,q_2,i_3,\ldots,r_\infty+1)$ of O_2 is contained in a $\mathcal{T}_2$-class. The collection of elements of O_1 which are mapped to $(i_1,q_2,i_3,\ldots,r_\infty+1)\mathcal{T}_2$ by $\mathscr{X}^{1,2}$ are all vectors of the form $(-,-,-,i_3,q_4,\ldots,r_\infty)$. This subset of O_1 is an $\mathcal{S}_3$-class. Again, $(\tilde{s}\mathcal{S}_3)\mathcal{X}_{1,2}$ is a collection of supremums of $\mathcal{T}_1$-classes. By taking the union of these $\mathcal{T}_1$-classes we obtain all vectors of O_2 of the form $(-,-,-,-,i_3,q_4,\ldots,r_\infty)$. It now follows that for (f,g) in $A(\Omega)$,
$f_{3,(-,-,-,i_3,q_4,\ldots,r_\infty)} = g_{2,(-,-,i_3,q_4,\ldots,r_\infty+1)} = g_{4,(-,-,-,-,i_3,q_4,\ldots,r_\infty)}$.
By continuing this procedure, we conclude that for $\gamma \in \Gamma$, when $\gamma \equiv 1 \pmod 2$,
$f_{\gamma,(q_0,i_1,q_2,i_3,\ldots,r_\infty)} = g_{\gamma-1,(i_1,q_2,i_3,q_4,\ldots,r_\infty+1)} = g_{\gamma+1,(-,-,i_1,q_2,i_3,\ldots r_\infty)}$.

The point $(i_1,q_2,i_3,q_4,\ldots,r_\infty+1)$ is also contained in a $\mathcal{T}_1$-class. All elements of $\tilde{s}\mathcal{S}_2$ get mapped to this $\mathcal{T}_1$-class by $\mathscr{X}^{1,2}$. We map this $\mathcal{S}_2$-class to O_2 and take the union of the corresponding $\mathcal{T}_1$-classes. We thus obtain a $\mathcal{T}_3$-class. This procedure can again be continued, and we may conclude that for $\gamma \geq 2$, $\gamma \in \Gamma$, if $\gamma \equiv 0 \pmod 2$ then $f_{\gamma,(q_0,i_1,q_2,i_3,\ldots,r_\infty)} =$

$g_{\gamma-1,(i_1,q_2,i_3,\ldots,r_\infty+1)} = g_{\gamma+1,(-,-,i_1,q_2,i_3,\ldots,r_\infty)}$.

We may now start with the vectors of O_2 and use the maps $\mathscr{X}^{2,1}$ and $\mathcal{X}_{2,1}$. By similar arguments to the ones above, we see that for (f,g) in $A(\Omega)$ and γ in $\Gamma \setminus \{0,\infty\}$, $f_{\gamma-1,(q_1,i_2,q_3,i_4,\ldots,r_\infty)} = f_{\gamma+1,(-,-,q_1,i_2,q_3,i_4,\ldots,r_\infty-1)} =$

$g_{\gamma,(i_0,q_1,i_2,q_3,i_4,\ldots,r_\infty)}$.

The above arguments set up a correspondence between the $\mathcal{S}_\infty$-class which consists of vectors with ∞-coordinate r_∞ and the $\mathcal{S}_\infty$-class which consists of vectors with ∞-coordinate $r_\infty-1$. There is also a correspondence between the $\mathcal{T}_\infty$-class which contains vectors with ∞-coordinate r_∞ and the $\mathcal{T}_\infty$-class which contains vectors with ∞-coordinate $r_\infty+1$.

Let P_1 denote the projection of elements of $A(\Omega)$ onto their first coordinates and let P_2 denote the projection of elements of $A(\Omega)$ onto their second coordinates. By using the above analysis, the reader may verify the components of (P_1,O_1) are as follows: (1) if $\gamma \in \Gamma\backslash\{\infty\}$ and $\gamma \equiv 0 \pmod 2$ then the γ-th component of (P_1,O_1) is ℓ-isomorphic to $(A(Q),Q)$; (2) if $\gamma \in \Gamma\backslash\{0,\infty\}$ and $\gamma \equiv 1 \pmod 2$ then the γ-th component of (P_1,O_1) is ℓ-isomorphic to $(A(\mathbb{I}),\mathbb{I})$; and (3) for $\gamma = \infty$, the γ-th component of (P_1,O_1) is ℓ-isomorphic to $(H_1,\mathbb{R})$ where $H_1 = \{h \in A(\mathbb{R}) \mid h(-1) = (-1)h\}$. We may similarly specify the components of (P_2,O_2). It now follows that P_1 is transitive on O_1 and P_2 is transitive on O_2.

We thus conclude that the orbits of $A(\Omega)$ in Ω are precisely O_1 and O_2. Hence, $A(\Omega)$ is an ℓ-group with the disjoint support property. The desired congruence structure on O_1 and O_2 follows from the properties of the maps.

The po-sets which are basic to the understanding of this theory are those such that $A(\Omega)$ has two orbits in Ω. However, many interesting examples can be constructed if we allow more than two orbits. We now give an example of a po-set Ω such that $A(\Omega)$ is an ℓ-group with the disjoint support property, and $A(\Omega)$ has an infinite number of orbits in Ω. The notation used previously should be generalized in the obvious way.

EXAMPLE 8. Let $\Gamma = \{0,1,2,\ldots\}$ be an index set, and, for each γ in Γ, let Q_γ be a copy of the rational numbers. We wish to define a partial order of the set $\Omega = \dot{\cup}\{Q_\gamma \mid \gamma \in \Gamma\}$. The most economical way to describe the relation is to draw a diagram and then give a brief explanation:

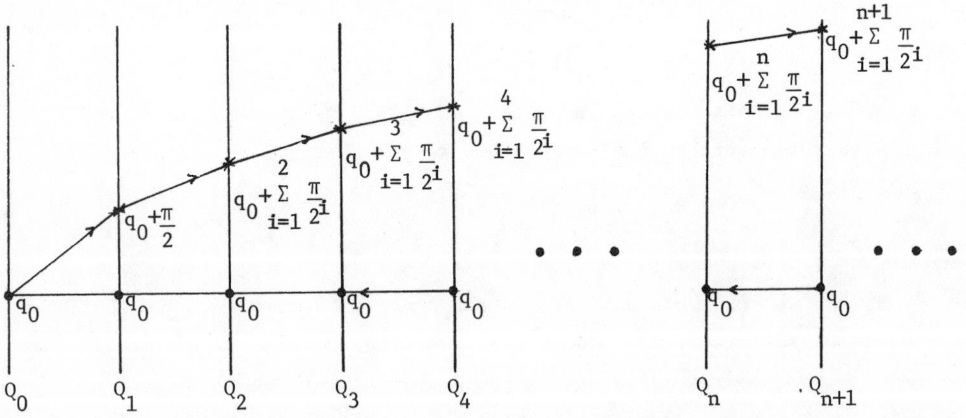

The diagram should suggest the following:

(1) For m,n in Γ where $m < n$, $\mathcal{X}_{m,n}$ (which is the map which defines

the sup of the lower bounds) is the identity map. We wish $\mathfrak{X}^{n,m}$, the map which defines the inf of the upper bounds, to be the inverse of $\mathfrak{X}_{m,n}$ so $\mathfrak{X}^{n,m}$ is the identity map.

(2) Let m = 0 and $n \in \Gamma$, $n \geq 1$. For q in Q_0, we define $\mathfrak{X}^{0,n}$ by $q\mathfrak{X}^{0,n} = q + \sum_{i=1}^{n} \pi/2^i$.

(3) We now use (2) to define $\mathfrak{X}^{m,n}$ and $\mathfrak{X}_{n,m}$ for m,n in Γ where m < n. The relation which should hold is $\bar{\mathfrak{X}}^{0,n} = \bar{\mathfrak{X}}^{0,m}\bar{\mathfrak{X}}^{m,n}$, i.e., $\bar{\mathfrak{X}}^{m,n} = (\bar{\mathfrak{X}}^{0,m})^{-1}\bar{\mathfrak{X}}^{0,n}$, i.e., $\bar{\mathfrak{X}}^{m,n} = \bar{\mathfrak{X}}_{m,0}\bar{\mathfrak{X}}^{0,n}$, i.e., $\bar{\mathfrak{X}}^{m,n} = (-\sum_{i=1}^{m} \pi/2^i)(+\sum_{i=1}^{n} \pi/2^i)$. $\bar{\mathfrak{X}}_{n,m} = (\bar{\mathfrak{X}}^{m,n})^{-1}$.

Note that points of Q_0 are covered by no points of Ω, each point of Q_1 is covered by one point of Ω, each point of Q_2 is covered by two points of Ω, etc. Thus, elements of A(Ω) must map each chain back to itself, and we may therefore consider A(Ω) as a subgroup of $\prod_{\gamma\in\Gamma} A(Q_\gamma)$. Let $G = \{g \in \prod_{\gamma\in\Gamma} A(Q_\gamma) \mid$ for each pair γ_1, γ_2 in Γ, $g(\gamma_1) = g(\gamma_2)$ and there exists m/n in Q such that $g(\gamma) = +m/n\}$. A straightforward computation shows each element of G is order preserving on Ω so $G \leq A(\Omega)$. Under the natural action, G is transitive on each of the chains Q_γ, so it follows that the collection of orbits of A(Ω) in Ω is $\{Q_\gamma \mid \gamma \in \Gamma\}$. Thus A(Ω) is an ℓ-group which has the disjoint support property.

We proceed to analyze the po-set Ω and the ℓ-group A(Ω) further. By restricting the order relations on Ω to $Q_0 \dot{\cup} Q_1$, we obtain a po-set of the type examined in Example 6. Using the analysis given there, $A(\Omega_{0,1}) = \{(f,f) \in A(Q) \times A(Q) \mid f(+\pi/2) = (+\pi/2)\}$. If we project the elements of $A(\Omega_{0,1})$ onto their first coordinates, we obtain an ℓ-group P_1 of A(Q) which contains all permutations which are translations by a particular rational number as well as many more permutations. To see that P_1 contains many more functions than rational translations, the reader is referred to [3]. In fact, if we restrict the order relations of Ω to any finite subcollection of the chains $\{Q_\gamma \mid \gamma \in \Gamma\}$ to form a po-set Σ, A(Σ) will be ℓ-isomorphic to an ℓ-subgroup of the direct product of a finite number of the ℓ-groups $A(Q_\gamma)$. Each element of A(Σ) will have the same permutation from A(Q) appearing in each of its coordinates, and all the members of A(Q) which may appear as the coordinate of some member of A(Σ) must commute with a particular irrational number. The ℓ-group formed by projecting the elements of A(Σ) onto one of their coordinates will again contain all rational translations as well as many more permutations.

However, when we compute the ℓ-group A(Ω), we see that the ℓ-group constructed by projecting the elements of A(Ω) onto a fixed coordinate consists only of the rational translations. This statement follows from the following facts. Let P_γ denote the projection of elements of $A(\Omega)$ onto their γ-th coordinates. Consider the additive subgroup H of $\mathbb{R}$ generated by $\{\sum_{i=1}^{n} \pi/2^i | n \in \Gamma\}$. H is a dense subgroup of $\mathbb{R}$, and each element of P_γ must commute with each element of H to preserve order on Ω. The only elements of $A(Q)$ which commute with translation by all numbers which are elements of a dense subset of $\mathbb{R}$ are the permutations which are rational translations.

Thus, we have given an example of a po-set Ω such that if the order relation on Ω is restricted to a finite subcollection of the orbits of $A(\Omega)$ in Ω, we obtain a po-set Σ such that $A(\Sigma)$ is an ℓ-group which is "highly transitive" on each of its orbits in Σ. But, if we consider all order relations on Ω, $A(\Omega)$ is "uniquely transitive" on each of its orbits in Ω.

Besides the disjoint support property and being laterally complete, the other characteristics of the ℓ-group of order automorphisms of a chain we mentioned were the properties of having closed stabilizers and being completely distributive. As we have said previously, the po-sets and ℓ-groups which are of primary importance in understanding this theory are those such that $A(\Omega)$ has two orbits in Ω. It can be shown, through a fairly lengthy case argument, that if Ω is a po-set such that $A(\Omega)$ is an ℓ-group having the disjoint support property and $A(\Omega)$ has two orbits in Ω, then $A(\Omega)$ has closed stabilizers and $A(\Omega)$ is completely distributive. We give no proof of these facts here, but will demonstrate how the arguments go by considering the specific examples we have just presented.

Therefore suppose Ω and $A(\Omega)$ satisfy the hypotheses of the previous paragraph. Let O_1 and O_2 denote the orbits of $A(\Omega)$ in Ω. $A(\Omega)$ is an ℓ-subgroup of $A(O_1) \times A(O_2)$. Let P_1 denote the projection of elements of $A(\Omega)$ onto their first coordinates and let P_2 denote the projection of elements of $A(\Omega)$ onto their second coordinates. Then, the following proposition is not hard to verify.

PROPOSITION 9. (1) If (P_1,O_1) and (P_2,O_2) have closed stabilizers, then $A(\Omega)$ has closed stabilizers.

(2) If P_1 and P_2 are completely distributive ℓ-groups, then $A(\Omega)$ is a completely distributive ℓ-group.

In [8], S. H. McCleary showed that if G is a transitive ℓ-subgroup of A(S), where S is a chain, and G has closed stabilizers, then G is completely distributive. Thus, if we show (P_1,O_1) and (P_2,O_2) have closed stabilizers, then we have shown (1) $A(\Omega)$ has closed stabilizers, (2) P_1 and P_2 are completely distributive, and $A(\Omega)$ is completely distributive.

THEOREM 10. The ℓ-groups $A(\Omega)$ studied in Examples 3,4,5,6 have closed stabilizers and are completely distributive.

Proof. In Examples 3,4,5, $P_1 = A(O_1)$ and $P_2 = A(O_2)$. For any chain S, A(S) always has closed stabilizers and is completely distributive. Hence, in the first three cases, the theorem follows. In [7], S. H. McCleary shows that any periodically o-primitive ℓ-permutation group has closed stabilizers. In Example 6, P_1 and P_2 are periodically o-primitive ℓ-permutation groups. Thus, the statements in the theorem hold.

The only example left to check is Example 7. By checking the classification given in [1], we see that the groups P_1 and P_2 of Example 7 are what we called "generalized periodic groups". The reason for this terminology is the following. Let S be a chain and $z: S \to \overline{S}$ be order preserving and be such that $\{sz^n | n = 0,1,2,\ldots\}$ is cofinal in $\overline{S}$. Then z can be extended to an order preserving function $\overline{z}: \overline{S} \to \overline{S}$, and we define $\mathcal{P}_z(S) = \{f \in A(S) | fz = zf\}$, a generalized periodic group. To obtain one of the periodic groups introduced by McCleary, we further require that z be one-to-one and Sz be dense in $\overline{S}$. In Example 7, for P_1, $z_1 = \mathcal{X}^{1,2}\mathcal{X}^{2,1}$, and for P_2, $z_2 = \mathcal{X}^{2,1}\mathcal{X}^{1,2}$. z_1 and z_2 are not one-to-one in these cases, but O_1z_1 is dense in $\overline{O}_1$ and O_2z_2 is dense in $\overline{O}_2$. Off-hand it may not be clear to the reader that the ℓ-groups P_1 and P_2 introduced in Example 7 are equal to $\mathcal{P}_{z_1}(O_1)$ and $\mathcal{P}_{z_2}(O_2)$ respectively. To see this, we will make a general study of ℓ-groups $\mathcal{P}_z(S)$ where z is not one-to-one but Sz is dense in S. Once it is obvious that P_1 and P_2 are of this type, we will show the ℓ-groups $\mathcal{P}_z(S)$ have closed stabilizers, and thus are completely distributive.

Therefore, suppose S is a homogeneous chain, z is an order preserving map from S into $\overline{S}$, for any s in S $\{sz^n | n = 0,1,2,\ldots\}$ is cofinal in $\overline{S}$, Sz is dense in $\overline{S}$, and z is not one-to-one. Further suppose $\mathcal{P}_z(S) = \{f \in A(S) | fz=zf\}$ is transitive on S. $\mathcal{P}_z(S)$ is an ℓ-subgroup of A(S). Define a convex equivalence relation Ker(z) on S by: for s_1,s_2 in S, s_1 Ker(z) s_2 if and only if $s_1z = s_2z$. Then Ker(z) is a convex $\mathcal{P}_z(S)$-congruence on S. We will now look closer at the chain Sz.

PROPOSITION 11. $\{\bar{f} | f \in \mathcal{P}_z(S)\}$ is transitive on Sz.

Proof. Let s_1z, $s_2z \in Sz$. Then $s_1, s_2 \in S$. $\mathcal{P}_z(S)$ is transitive on S so there exists f in $\mathcal{P}_z(S)$ such that $s_1f = s_2$. Therefore $s_1fz = s_2z$, i.e., $s_1zf = s_2z$.

Since Sz is dense in $\bar{S}$, we may ℓ-embed $\mathcal{P}_z(S)$ as a transitive ℓ-subgroup of A(Sz) (Note this does not mean that $\mathcal{P}_z(S) = \mathcal{P}_z(Sz)$). Since z is not one-to-one, Ker(z) = $\mathcal{K}_1$ is a proper convex $\mathcal{P}_z(S)$-congruence on S. The following proposition shows that $\mathcal{K}_1$ can be thought of as a convex congruence on Sz.

PROPOSITION 12. For each s in S, sz is an element of the convexification of some $\mathcal{K}_1$-class, i.e., sz is not an endpoint of a $\mathcal{K}_1$-class or a hole outside all of the classes.

Proof. This follows from the denseness of Sz in S and the transitivity of $\mathcal{P}_z(S)$ on Sz.

We now start drawing conclusions about the chain S.

PROPOSITION 13. There is a countable tower of proper convex $\mathcal{P}_z(S)$-congruences on S.

Proof. It suffices to show that each point of S is contained in a countable tower of o-blocks. Choose s_0 in S. $s_0 \in s_0\mathcal{K}_1$. $(s_0\mathcal{K}_1)z = s_0z$. By Proposition 12, s_0z is an element of the convexification of some $\mathcal{K}_1$-class, call it $s_1\mathcal{K}_1$. Let $K_1 = \{sz | sz$ is an element of the convexification of $s_1\mathcal{K}_1\}$. Since Sz is dense in $\bar{S}$, this is an infinite collection of points of $\bar{S}$, and it contains s_0z. Let $s_0\mathcal{K}_2 = \{s \in S | sz \in K_1\}$. $s_0\mathcal{K}_2 \supset s_0\mathcal{K}_1$, and $s_0\mathcal{K}_2$ is an o-block for a unique convex $\mathcal{P}_z(S)$-congruence $\mathcal{K}_2$ which contains $\mathcal{K}_1$ properly. s_0z is an element of the convexification of a $\mathcal{K}_2$-class. Let $K_2 = \{sz | sz$ is an element of the convexification of $s_1\mathcal{K}_2\}$. We can use K_2 to transfer back to s_0 obtaining a convex $\mathcal{P}_z(S)$-congruence $\mathcal{K}_3$ such that $\mathcal{K}_3 \supset \mathcal{K}_2 \supset \mathcal{K}_1$. We may continue this process countably many times.

PROPOSITION 14. For each s_0 in S, the tower of o-blocks about s does not intersect the tower of o-blocks about sz.

Proof. The tower of congruences is $\mathcal{K}_1 \subset \mathcal{K}_2 \subset \mathcal{K}_3 \subset \dots$. Suppose for some n, s_0 and s_0z are elements of the convexification of $s_0\mathcal{K}_n$. As we have seen previously, it makes sense to talk about $s_0z\mathcal{K}_i$ for i = 1,2,... . So, $s_0\mathcal{K}_n \supset s_0z\mathcal{K}_{n-1}$. If we fix $s_0\mathcal{K}_n$, we must fix $s_0z\mathcal{K}_{n-1}$ (by the way the correspondence of z works), so, for each s in $s_0z\mathcal{K}_{n-1}$, s cannot be mapped to any point in $s_0\mathcal{K}_n \backslash s_0z\mathcal{K}_{n-1}$. This contradicts the transitivity assumption on $\mathcal{P}_z(S)$.

It follows from this that the join of the o-blocks containing any one point is a proper o-block. Thus $\bigvee_{i=1}^{\infty} \mathcal{K}_i = \mathcal{K}_\infty$, a proper convex $\mathcal{P}_z(S)$-congruence.

We can now discuss the ℓ-group $\mathcal{P}_z(S)$. From the above discussion, $\mathcal{P}_z(S)$ can be ℓ-embedded in a wreath product with countably many factors where the index set on the wreath product has both a smallest and a largest element. We denote this by $\Gamma = \{1,2,3,4,\ldots,\infty\}$. The z function, at the ∞-level, gives a periodic correspondence like the ones studied by McCleary by matching $s\mathcal{K}_\infty$ with $sz\mathcal{K}_\infty$. Since classes are matched to classes, we have a periodic ℓ-permutation group of Config(1), and, thus, at the ∞-level we are periodically o-primitive [6]. Call the ∞-component $(\mathcal{P}_z(S_\infty),S_\infty)$.

Obviously, we can set up an order-isomorphism between Sz, the chain of $\mathcal{K}_1$-classes, the chain of $\mathcal{K}_2$-classes, etc. At the local level, when we choose functions for our component from $A(s\mathcal{K}_1)$, we must be sure that the functions map the points from Sz which lie in that class back to themselves. This is the only restriction. Let $sz\mathcal{K}_1$ denote $\{sz \in Sz \mid sz$ is an element of the convexification of $s\mathcal{K}_1\}$. Then, locally, the ℓ-group we induce is $A(s\mathcal{K}_1,sz\mathcal{K}_1) = \{f \in A(s\mathcal{K}_1) \mid (sz\mathcal{K}_1)\bar{f} = sz\mathcal{K}_1\}$. Therefore, out local component is $(A(s\mathcal{K}_1,sz\mathcal{K}_1),s\mathcal{K}_1)$. For all levels in the wreath product except the ∞-level, the ℓ-group part of the component will be ℓ-isomorphic to $A(s\mathcal{K}_1,sz\mathcal{K}_1)$ because of the correspondence set up by z, but the remaining chains will be ℓ-isomorphic to $sz\mathcal{K}_1$ because of the first sentence of this paragraph. Thus, $\mathcal{P}_z(S) \subseteq (A(s\mathcal{K}_1,sz\mathcal{K}_1),s\mathcal{K}_1)\mathrm{Wr}\,(A(s\mathcal{K}_1,sz\mathcal{K}_1),sz\mathcal{K}_1)\mathrm{Wr}\,(A(s\mathcal{K}_1,sz\mathcal{K}_1),sz\mathcal{K}_1)\mathrm{Wr}\ldots \mathrm{Wr}(\mathcal{P}_z(S_\infty),S_\infty)$, the top ℓ-group is periodically o-primitive, and the only other restrictions are just the tie-ins at different levels induced by z.

We now see that the ℓ-groups P_1 and P_2 of Example 7 are of this type. For P_1, $s\mathcal{K}_1 = sz\mathcal{K}_1 = Q \overleftarrow{\times} \mathbb{I}$, and for P_2, $s\mathcal{K}_1 = sz\mathcal{K}_1 = \mathbb{I} \overleftarrow{\times} Q$. In both cases $(\mathcal{P}_z(S_\infty),S_\infty) = (G,\mathbb{R})$ where $G = \{f \in A(\mathbb{R}) \mid f(+1) = (+1)f\}$. We are now in a position to show these ℓ-groups have closed stabilizers.

THEOREM 15. Suppose S is a homogeneous chain, $z:S \to \bar{S}$ is order preserving, for each s in S $\{sz^n \mid n = 0,1,2,\ldots\}$ is cofinal in $\bar{S}$, z is not one-to-one, but

Sz is dense in $\overline{S}$. Further suppose $\mathcal{P}_z(S) = \{f \in A(S) | fz = zf\}$ is transitive on S. Then $\mathcal{P}_z(S)$ has closed stabilizers.

Proof. As shown previously, $\mathcal{P}_z(S)$ can be ℓ-embedded in A(Sz) by first extending the functions in $\mathcal{P}_z(S)$ to $\overline{S}$ and cutting them back to Sz. Call this new ℓ-group G. To prove the theorem, it suffices to show (G,Sz) has closed stabilizers, since if G has closed stabilizers at points of Sz it will have closed stabilizers at points of S [5, Theorem 7].

Let $\{g_i | i \in I\} \subseteq G_{s_0z}$ and suppose $\bigvee_i g_i = g$ exists. By way of contradiction suppose $s_0z < s_0zg$. We wish to construct h such that $h \geq g_i$ for all i and $h < g$. In Sz, we have $\{s_0z^n | n = 1,2,\ldots\}$, all of which are contained in non-intersecting towers of $\mathcal{K}_i$-classes. We also have the $\mathcal{K}_1$-class $s_0\mathcal{K}_1$ to the left of s_0z which is mapped to s_0z by z, a $\mathcal{K}_2$-class to the left of $s_0\mathcal{K}_1$ which is mapped to $s_0\mathcal{K}_1$ by z, etc. Outside of the towers surrounding all of these classes, let h = g. Since g moves s_0z up, it must move $s_0\mathcal{K}_1$ up to $s_0g\mathcal{K}_1$. We let h move $s_0\mathcal{K}_1$ to $s_0g\mathcal{K}_1$ (and hence the $\mathcal{K}_2, \mathcal{K}_3, \ldots$ classes connected with this by z to the same classes), but we will alter the action of g inside $s_0\mathcal{K}_1$. We can define h so that for some s'z in $s_0\mathcal{K}_1$, $s'zh < s'zg$ and for all other points sz in $s_0\mathcal{K}_1$, $szh \leq szg$. This makes $h < g$, but, by still moving up $s_0\mathcal{K}_1$ and the corresponding classes, we have $h \geq g_i$ for all i, a contradiction. Thus, G has closed stabilizers.

It now follows from the theorem that the ℓ-groups P_1 and P_2 associated with Example 7 have closed stabilizers, and hence are completely distributive. Thus, the ℓ-group A(Ω) from Example 7 has closed stabilizers and is completely distributive. This completes the verification of what we claimed held true for the examples presented.

REFERENCES

1. M. A. Bardwell, Lattice-ordered groups of order automorphisms of partially ordered sets. *Pre-print*, (1978).

2. A. M. W. Glass, *Ordered Permutation Groups*, Bowling Green State University, Bowling Green, (1976).

3. W. C. Holland, The lattice-ordered group of automorphisms of an ordered set. *Mich. Math. J. 10:* 399-408 (1963).

4. W. C. Holland, Transitive lattice-ordered permutation groups. *Math. Zeit. 87:* 420-433 (1965).

5. S. H. McCleary, Closed subgroups of lattice-ordered permutation groups. *Trans. A.M.S. 173:* 303-314 (1972).

6. S. H. McCleary, O-primitive ordered permutation groups. *Pacific J. Math 40:* 349-372 (1972).

7. S. H. McCleary, O-primitive ordered permutation groups II. *Pacific J. Math 49:* 431-441 (1973).

8. S. H. McCleary, Pointwise suprema of order preserving permutations. *Illinois J. Math. 16:* 69-75 (1972).

AUTOMORPHISM GROUPS OF MINIMAL η_α-SETS

Elliot C. Weinberg
University of Illinois at Urbana-Champaign
Urbana, Illinois

The lattice-ordered groups (ℓ-groups) of order-automorphisms of chains are universal for ℓ-groups [4]. The η_α-sets are universal for chains [3]. Hence every ℓ-group can be embedded in the ℓ-group of o-automorphisms of an η_α-set. Capitalizing on this fact and the assumption (a version of the generalized continuum hypothesis) that there exist, for arbitrarily large α, η_α-sets of cardinal $\aleph_\alpha$, a number of interesting embedding theorems for ℓ-groups have been proved: every ℓ-group is embeddable in a divisible ℓ-group [4,8], in an ℓ-group all of whose ℓ-automorphisms are inner [5], in an ℓ-group in which any two strictly positive elements are conjugate [6]; every archimedean totally ordered group is amalgamable in the variety of all lattice-ordered groups [6]. A unified discussion appears in [1].

Harzheim [2] has produced a theory of "minimal" η_α-sets based on order-type rather than cardinality. The purpose of this paper is to investigate the ℓ-group of o-automorphisms of a minimal η_α-set and indicate how it can be used to obtain the various embedding theorems for ℓ-groups without any dependence on the generalized continuum hypothesis.

1. TAME EMBEDDINGS IN η_α-SETS

DEFINITION 1.1. A subset S of a chain (totally ordered set) T is said to be *tamely embedded* in T if every o-automorphism (order-automorphism) of S extends to an o-automorphism of T.

For example, every chain is tamely embedded in its Dedekind completion and in its chain of ideals. If T is the chain Q of rational numbers with one gap filled, then Q is not tamely embedded in T although it can be reimbedded tamely. Notice also that Q cannot be tamely embedded in the irrationals although the two chains have ℓ-isomorphic ℓ-groups of o-automorphisms. On the other hand any chain that can be embedded in the real numbers can be tamely embedded there.

The ℓ-group of all o-automorphisms of a chain T will be denoted A(T). In this paper o-automorphisms will be written as acting on the left.

LEMMA 1.2 [8]. If S is tamely embedded in the chain T, then the ℓ-group A(S) can be embedded in the ℓ-group A(T).

The embedding of A(S) in A(T) obtained in the proof of the lemma is by extension of o-automorphisms; i.e., if ψ denotes the embedding, $\psi(f)|S = f$ for each f in A(S).

DEFINITION 1.3. Let α be a fixed ordinal number. Call a set *small* if it has cardinal less than $\aleph_\alpha$. Thus a cardinal is regular if the union of a small number of small sets is small. Henceforth we will assume that $\aleph_\alpha$ is regular. A chain is called *order-small* if every subset which is well-ordered or dually well-ordered is small. (Harzheim calls such chains ω_α-free.) A chain is called *almost-order-small* if it is the union of a family, indexed by ω_α, of order-small subsets. We can always assume that the family is increasing.

If $\alpha = 0$, then small and order-small mean finite, while almost-order-small means countable. If $\alpha = 1$, then small means countable, while the reals, for example, are order-small, but not small.

PROPOSITION 1.4. (a) Any subset of an order-small chain is order-small.

(b) The lexicographic product of two order-small chains is order-small.

(c) The lexicographic union of a family of order-small chains over an order-small index set is order-small.

(d) Each of the above is true for almost-order-small sets.

DEFINITION 1.5. A subset S of a chain T *splits small sets* in T if, whenever $A \cup B$ is a small subset of T and $A < B$, there exists s in S such that $A < s < B$. A chain is an η_α*-set* if it has a subset which splits small sets. A chain in which every bounded open interval is an η_α-set is a *central*-η_α*-set*.

In the category of chains and o-monomorphisms, the η_α-sets are the objects which are injective relative to order-small chains (equivalently, relative to small chains) [9]. Hence any almost-order-small set is embeddable in any η_α-set. For this reason an almost-order-small η_α-set will be called *minimal*. Harzheim [2] has proved that there exists a minimal η_α-set Q_α and that any two minimal η_α-sets are o-isomorphic. The chain Q_α is transitive and o-2-transitive. Its cardinal is

$$\aleph_\alpha^* = \Sigma_{k < \aleph_\alpha} 2^k.$$

In order to obtain *tame* embeddings of almost-order-small chains in Q_α, Harzheim's construction is modified below.

DEFINITION 1.6. An ideal I of a chain T will be called a *small ideal* if it has a small cofinal subset. Let I(T) and $I_\alpha(T)$ denote the sets of ideals and small ideals of T, respectively. Extend the order on T to

$$T^+ = T \cup I(T) \text{ and}$$

$$T^{(\alpha)} = T \cup I_\alpha(T)$$

by

$$t < I \text{ if } t \in I.$$

Define the chains $T_{(i)}$ inductively by

$$T_{(0)} = T,$$

$$T_{(i+1)} = T_{(i)}{}^{(\alpha)}, \text{ and}$$

$$T_{(j)} = \cup_{i<j} T_{(i)}$$

for j a limit ordinal. The chains ${}_+T_{(i)}$ are defined similarly in terms of the embedding $T \to T^+$.

PROPOSITION 1.7. Let T be a chain. Then T is tamely embedded in T^+. If T is order-small, then so is T^+. If T is almost-order-small, then so is $T^{(\alpha)}$.

Proof. The tameness of the embeddings is clear.

Suppose that T is the union of a nondecreasing ω_α-sequence $(T_j)_{j < \omega_\alpha}$ of order-small chains. Let I_i be the set of all small ideals which have a cofinal set in T_i. If I is any ideal with a small cofinal set C, then, by the regularity of the cardinal, there exists $j < \omega_\alpha$ such that C is contained in T_j, hence $I(T) = \bigcup_{j < \omega_\alpha} I_j$. Suppose that $(I_j)_{j < \omega_\alpha}$ were a strictly increasing ω_α-sequence of ideals in I_i. For each $j < \omega_\alpha$ there would exist an x_j which is in $I_{j+1} \cap T_i$ but not in I_j, so there would exist a strictly increasing ω_α-sequence (x_j) in T_i. By a similar argument, I_i has no strictly decreasing ω_α-sequence. Thus I_i and $T_i \cup I_i$ are order-small, so $T^{(\alpha)} = \bigcup_{i < \omega_\alpha} (T_i \cup I_i)$ is almost-order-small.

Clearly, T^+ is order-small when T is order-small.

Harzheim achieves the construction of Q_α by using 'decompositions' of T, partitions (A,B) of T into an ideal and a dual ideal. This yields the same chain T^+ of course. We could also use 'small decompositions', decompositions (A,B) such that A has a small cofinal subset and B a small coinitial subset, to construct an analogue of $T^{(\alpha)}$. Then 1.7 would still be valid. Notice that a chain is an η_α-set exactly when it has no small decompositions, so that we would have, in this case, $T = T^{(\alpha)}$ exactly when T is an η_α-set.

THEOREM 1.8. Any almost-order-small chain can be tamely embedded in the minimal η_α-set Q_α.

Proof. If T is almost-order-small, then, by induction, $T_{(\omega_\alpha)}$ is almost-order-small and contains a tamely embedded copy of T. If $A \cup B$ is a small subset of $T_{(\omega_\alpha)}$, then, for some $j < \omega_\alpha$, $A \cup B$ is contained in $T_{(j)}$. If I is the ideal in $T_{(j)}$ generated by A, then $A < I < B$ in $T_{(j+1)}$. Hence $T_{(\omega_\alpha)}$ is a minimal η_α-set.

2. THE SUB-ℓ-GROUPS OF $A(Q_\alpha)$

DEFINITION 2.1. A family of prime subgroups of an ℓ-group is called *dense* if its intersection is the identity; the family is called *representing* if the intersection contains no nontrivial ℓ-ideal. We say that a prime subgroup P of G has a property *residually* if the chain G/P has the property.

Any representing family P induces an embedding in the product of the ℓ-groups A(G/P), P in P. Let $\leq$ be any total order of P, and let $T = \bigcup_{P \in P} G/P$ be the lexicographic union: $gP_1 \leq hP_2$ exactly when $P_1 < P_2$, or $P_1 = P_2$ and $gP_1 \leq hP_1$. This induces a natural embedding of G in A(T): $g \to \bar{g}$ where $\bar{g}(hP) = ghP$ [4].

THEOREM 2.2. Let G be an ℓ-group. The following are equivalent.

(a) G is embeddable in $A(Q_\alpha)$.

(b) G has a dense family of at most $\aleph_\alpha^*$ prime subgroups which are residually almost-order-small.

(c) G has a representing family of at most $\aleph_\alpha^*$ prime subgroups which are residually almost-order-small.

Proof. Assume that G is a sub-ℓ-group of $A(Q_\alpha)$. For each t in Q_α, consider the stabilizer subgroup G_t of elements of G which fix t. There are at most $\aleph_\alpha^*$ of these subgroups, each is prime, their intersection is the identity, and each chain G/G_t, being o-isomorphic to the subchain Gt of Q_α, is almost-order-small. Thus (b) holds. Clearly, (b) implies (c).

To see that (c) implies (a), let P be the given family of prime subgroups. Since Q_α has cardinal $\aleph_\alpha^*$, we may give P an almost-order-small total order. The lexicographic union $T = \bigcup_{P \in P} G/P$ is almost-order-small and hence tamely embeddable in Q_α. Then G is embeddable in A(T) and hence in $A(Q_\alpha)$.

COROLLARY 2.3. The ℓ-group $A(Q_\alpha)$ contains every ℓ-group of cardinal not exceeding $\aleph_\alpha$, but, for any $\beta > \alpha$, there is an ℓ-group of cardinal $\aleph_\beta$ which is not embeddable in $A(Q_\alpha)$.

Proof. The universality of $A(Q_\alpha)$ follows from 2.2(b). For the counterexample consider the Hahn sum, over ω_β, of copies of the rationals.

COROLLARY 2.4. Every ℓ-group is embeddable in a divisible ℓ-group.

Proof. Since Q_α is doubly transitive, $A(Q_\alpha)$ is divisible.

Notice that the family of sub-ℓ-groups of $A(Q_\alpha)$ is closed under the formation of direct products of families with at most $\aleph_\alpha^*$ members.

THEOREM 2.5. Every ℓ-automorphism of $A(Q_\alpha)$ is inner. Every ℓ-group is embeddable in an ℓ-group all of whose ℓ-automorphisms are inner.

Proof. It suffices to prove that, for each x in the Dedekind completion of Q_α, the orbit $A(Q_\alpha)x$ is not o-isomorphic to Q_α. The chain $A(Q_\alpha)x$ is the set of points in the Dedekind completion which have the same character as x. If the right or left character of x is less than α, then $A(Q_\alpha)x$ is not an η_α-set. The set of points of character (α,α) has cardinal greater than $\aleph_\alpha^*$ [2].

It is also possible to carry out the various discussions given by McLeary in case there is an η_α-set of cardinal $\aleph_\alpha$ about orbits and gaps without making the cardinality assumption.

If we specialize the theorem characterizing sub-ℓ-groups of $A(Q_\alpha)$ to the case $\alpha = 0$, we get a description of the sub-ℓ-groups of $A(Q)$ in terms of the existence of an appropriate countable family of prime subgroups which are residually countable. There is an analogous description of the subgroups of $A(R)$, R the chain of real numbers, differing only in the condition that the prime subgroups residually have a countable order-dense subset. Details will appear elsewhere.

3. PARTITIONED CHAINS

(The material on partitioned chains and amalgamations will be presented in detail elsewhere.)

DEFINITION 3.1. Let f be in $A(T)$. A maximal interval of T which is left pointwise fixed by f is called a *zero interval* of f. If $f(t) > t$, then the convexification of $\{f^k(t)\}$ is called a *positive interval* of f.

Notice that the set of intervals of a positive element inherits an order from T which turns it into a chain, one which is naturally partitioned into two subchains. The theory of partitioned 2-chains behaves like the theory of chains. In the study of amalgamations it is convenient to have the analogue to minimal η_α-sets for 2-chains. In case $\aleph_\alpha = \aleph_\alpha^*$, the necessary material was obtained by Pierce [6].

DEFINITION 3.2. Let m be an ordinal number. An m-*chain* is an ordered pair

$$\mathcal{T} = (T,(T_i)_{i<m})$$

consisting of a chain T and a partition $(T_i)_{i<m}$ of T into subchains. An *o-homomorphism* $\phi: T \to S$ is an order-preserving function $\phi: T \to S$ of the chain T into the chain S such that ϕ maps T_i into S_i for each i. Call an o-homomorphism an *o-monomorphism* (resp., *o-isomorphism* if it is one-to-one (resp., one-to-one and onto). We say that T has an *order-theoretic property* if the underlying chain T has the property. *Sub-m-chains* are defined in the obvious way. Call T an η_α*-m-chain* if each T_i splits small sets of T. An η_α-m-chain is *minimal* if it can be embedded in any η_α-m-chain.

THEOREM 3.3. Let m be an ordinal whose cardinal does not exceed $\aleph_\alpha^*$.

If A is an order-small sub-m-chain of an almost-order-small m-chain B, then any o-monomorphism of A into an η_α-m-chain I can be extended to an o-monomorphism of B into I. In particular, any η_α-m-chain contains any almost-order-small η_α-m-chain.

There exists a minimal η_α-m-chain. An η_α-m-chain is minimal exactly when it is almost-order-small. Any two minimal η_α-m-chains are o-isomorphic.

4. AMALGAMATIONS

DEFINITION 4.1. An ℓ-group A is *amalgamable* if any set of embeddings of A can be realized by a single embedding; i.e., if ϕ_i: $A \to G_i$ are ℓ-monomorphisms, then there exists an ℓ-group B and ℓ-monomorphisms ψ_i: $G_i \to B$ such that $\psi_i\phi_i = \psi_j\phi_j$ for all i and j.

THEOREM 4.2. Every archimedean totally-ordered group is amalgamable.

The amalgamation is realized by choosing a regular cardinal $\aleph_\alpha$ with respect to which each G_i is small, and embedding G_i in $A(Q_\alpha)$ in such a way that the copies of A are conjugate. In order to achieve this, positive elements of the various copies of A must have o-isomorphic 2-chains of intervals such that corresponding intervals are o-isomorphic.

DEFINITION 4.3. A positive element of $A(Q_\alpha)$ is *broadly distributed* if its 2-chain of intervals is an η_α-2-chain and each zero interval is an η_α-set.

A broadly distributed element of $A(Q_\alpha)$ of necessity has each positive interval an almost-order-small central η_α-set with countable cofinal and coinitial subsets and has an almost-order-small interval 2-chain. Hence any two broadly distributed elements are conjugate in $A(Q_\alpha)$.

LEMMA 4.4. If the ℓ-group G is small with respect to the cardinal $\aleph_\alpha$, then G can be embedded in $A(Q_\alpha)$ in such a way that each strictly positive element of G is broadly distributed.

To achieve this, G is first embedded in A(T) for some order-small T, and T is tamely embedded in successively larger order-small chains alternately adding small ideals as in 1.7 so as to add elements between small sets and, by a more complicated construction introduced by Pierce, inserting new intervals both positive and zero between small sets of intervals of positive elements. After ω_α steps the embedding of the lemma is obtained. Besides the amalgamation theorem, the lemma has the following as a consequence.

THEOREM 4.5. Any ℓ-group can be embedded in an ℓ-group in which any pair of strictly positive elements is conjugate.

The same procedure which is used to prove that o-subgroups of the reals are amalgamable can be used to show that for other A, certain sets of embeddings of A can be realized by a single embedding. The following is related to work of Pierce [6] and Reilly [7].

THEOREM 4.6. Let A be a direct product of totally ordered archimedean groups. Suppose that A is embedded in various G_i in such a way that, for each strictly positive a in A and for each i, the orthogonal complement of a in G_i is normal in G_i. Then the embeddings may be realized in a single ℓ-group.

REFERENCES

1. A. M. W. Glass, *Ordered Permutation Groups*, Bowling Green State University, 1976.

2. E. Harzheim, Beiträge zur Theorie der Ordnungstypen, insbesondere der η_α-Mengen, *Math. Ann.* 154: 116-134 (1964).

3. F. Hausdorff, *Grundzüge der Mengenlehre,* Leipzig, 1914.

4. W. C. Holland, The lattice-ordered group of automorphisms of an ordered set, *Michigan Math. J.* 10: 399-408 (1963).

5. S. H. McLeary, The lattice-ordered group of automorphisms of an α-set, *Pacific J. Math.* 49: 417-424 (1973).

6. K. R. Pierce, Amalgamations of lattice-ordered groups, *J. Australian Math. Soc.* 13: 25-34 (1972).

7. N. R. Reilly, Permutational products of lattice-ordered groups, *J. Australian Math. Soc.* 13: 25-34 (1972).

8. E. C. Weinberg, Embedding in a divisible lattice-ordered group, *J. London Math. Soc.* 42: 504-506 (1967).

9. E. C. Weinberg, Relative injectives, *Symposia Mathematica*. 21: 555-564.

CUT COMPLETIONS OF LATTICE-ORDERED GROUPS BY CAUCHY CONSTRUCTIONS

Richard N. Ball
Boise State University
Boise, Idaho

The two classical constructions of the real numbers from the rational numbers are by Dedekind cuts and by Cauchy sequences. Each of these methods generalizes in several ways to arbitrary ℓ-groups; moreover, the interplay between them provides stronger results than would be possible by either method alone. It is the purpose of this talk to outline some ℓ-group results obtained by these methods and to raise questions whose solutions seem to be central to further progress.

Suppose we have a completeness property $\mathcal{P}$ for ℓ-groups. Familiar examples of such properties are lateral completeness, Dedekind-MacNeille completeness, and projectability. To say that an ℓ-group H is a $\mathcal{P}$-completion of an ℓ-group G means three things:

(1) G is large in H,

(2) H has property $\mathcal{P}$,

(3) H is minimal with respect to (2) in the sense that $G \leq K < H$ implies K lacks $\mathcal{P}$.

A completion result proves the existence and possibly even the uniqueness of a $\mathcal{P}$-completion for every ℓ-group in some class. Such results are intrinsically interesting since, in the case of an existence and uniqueness result, they challenge us to divine the structure of the completion, which is somehow implicit in G.

We are interested in two sorts of completeness properties. The first kind asserts the existence of a supremum in G for each subset of G of a given type. Examples are lateral and Dedekind-MacNeille completeness. To obtain the corresponding completion one would naturally attempt to adjoin to G suprema (and infima) for the desired subsets; this procedure is the method of cuts. The second sort of completeness is Cauchy completeness: the assertion that every Cauchy filter on G converges to a point of G. This technique requires a cogent definition of Cauchy filter and leads to a completion construction by Cauchy space techniques.

Cut completeness properties will be emphasized in this talk because they have the advantage of requiring no additional (i.e. Cauchy) structure in their definitions. Cut completion results, therefore, lie properly within the boundaries of ℓ-group theory and should be of interest to anyone professing that specialty. Cauchy constructions, on the other hand, require the imposition of additional (Cauchy) structure, and there may be objections to this technique on the grounds of complication. Such a point of view is difficult to defend, in my opinion, for several reasons. First of all, there are definitions of Cauchy filters, three to be mentioned shortly, which must seem quite natural to anyone familiar with ℓ-groups. Secondly, once having defined a Cauchy filter, the construction of a completion is quite elegant and not at all cumbersome. Third and most important, the completeness properties, both Cauchy completeness and cut completeness, of the Cauchy completions are striking. In more than one instance a new and interesting cut completeness property came from studying a Cauchy completion. Indeed, it is possible to take the point of view that Cauchy (and convergence) structures are tools without which it would be difficult to understand certain completeness properties of lattice ordered groups.

Not all completion results are accessible by these methods. In particular, both cut constructions and the Cauchy constructions we consider preserve the root structure of order closed primes. Therefore we cannot construct the projectable hull (or any of its variations) of a representable ℓ-group by these methods.

1. CUT COMPLETIONS

Suppose $G \leq H$ and that X_1 and X_2 are subsets of G such that $\bigvee X_i = h_i \in H$. The lattice and group theoretic properties of h_1 and h_2 are inherited from those of X_1 and X_2 in the sense that $h_1h_2 = \bigvee X_1X_2$, $h_1 \wedge h_2 = \bigvee(X_1 \wedge X_2)$ and $h_1 \vee h_2 = \bigvee(X_1 \vee X_2)$. The idea is to reconstruct h_1 and h_2 by adjoining to G the "cuts" X_1 and X_2.

To say that G is *completely embedded* in H, written $G \leqq H$, is to say that G is an ℓ-subgroup of H and that suprema in G and H agree. For $X \subseteq H$, $\mathrm{ocl}_H(X)$, the *order closure of X in H*, is defined inductively as follows. $X_0 = X$,

$X_{\alpha+1} = \{\bigvee S | S \subseteq X_\alpha\}$ for α even, $X_{\alpha+1} = \{\bigwedge S | S \subseteq X_\alpha\}$ for α odd, $X_\delta = \cup \{X_\alpha | \alpha < \delta\}$ for limit ordinals δ, and $\mathrm{ocl}_H(X) = X_\gamma$ for some ordinal γ such that $X_\gamma = X_{\gamma+1} = X_{\gamma+2}$. $X \subseteq H$ is *order closed* if $X = \mathrm{ocl}_H(X)$.

Given a lattice G, $X \subseteq G$ is an *element filter (element ideal)* if $y \geq x \in X$ ($y \leq x \in X$) implies $y \in X$ and if $x_1, x_2 \in X$ implies $x_1 \wedge x_2 \in X$ ($x_1 \vee x_2 \in X$). For example, if $G \leq H$ and $h \in H$ then $L_G(h) = \{f \in G | f \leq h\}$ is an element ideal and $U_G(h) = \{f \in G | f \geq h\}$ is an element filter. The unqualified term *filter (ideal)* refers to a filter of subsets of G (normal convex ℓ-subgroup of the ℓ-group G). For $X \subseteq G$ let $X' = \{y \in G | y = \bigvee(X \wedge y)\}$.

The first lemma, whose proof is a straightforward exercise in lattice theory, makes clear that we may as well assume cuts to be order closed element ideals.

LEMMA 1.1. Suppose $X \subseteq G \leq H$ and $\bigvee X = h \in H$. Then for $Y \subseteq G$ the following are equivalent:

(a) $Y = X'$,

(b) $Y = \cup \{Z \subseteq G | \bigvee Z = h\}$,

(c) $Y = L_G(h)$,

(d) $Y = \cap \{Z | Z$ is an order closed element ideal and $X \subseteq Z \subseteq G\}$.

When we adjoin suprema to G we are filling the holes in G. But, as the next result points out, some holes are too deep to be filled.

LEMMA 1.2. If $X \subseteq G \leq H$ and if $\bigvee X = h \in H$ then X' is not a union of cosets, right or left, of any non-trivial convex ℓ-subgroup of G. That is, $1 < c \in G$ implies $X'c \neq X'$ and $cX' \neq X'$.

Proof. If $X'c = X'$ then $hc = (\bigvee X')c = \bigvee(X'c) = \bigvee X' = h$, implying $c = 1$.

DEFINITION. $X \subseteq G$ is a *cut* if X is an order closed element ideal such that $1 < c \in G$ implies $Xc \neq X$ and $cX \neq X$. A cut X is a *flip flop cut* if $X = L(U(X))$. A cut X is a *locally finite cut* if $\cap \{(Xx^{-1} \vee 1)^{\perp\perp} | x \in X\} = \cap \{x^{-1}X \vee 1)^{\perp\perp} | x \in X\} = 1$. An ℓ-group G is *cut complete* if every cut of G has a supremum in G. Similarly, G is *flip flop cut complete (locally finite cut complete)* if every flip flop cut (locally finite cut) of G has a supremum in G.

G is *laterally complete* if for every disjoint $D \subseteq G$ the cut D' has a supremum in G. Notice that every such cut is locally finite. Holland, who coined the term "laterally complete", proved that every ℓ-group G can be ℓ-embedded in a laterally complete ℓ-group H. This result does not, however, provide a lateral completion since Holland's embedding does not guarantee that G is large in H. In [4], Conrad proved the existence and uniqueness of a lateral completion for every representable ℓ-group. The existence and uniqueness of a lateral completion for all ℓ-groups, due to Bernau [3], is a deep and important result in the theory of lattice ordered groups.

The difficulty of Bernau's proof might make us skeptical about the existence of cut completions or even about locally finite cut completions. It ought not daunt us. Rather, it should teach us to avoid a direct construction.

2. ℓ-CONVERGENCE AND ℓ-CAUCHY STRUCTURES

A definition of Cauchy filter involves, either explicitly or implicitly, a notion of convergence. To define a notion of convergence is to decide which filters $\mathcal{F}$ (of subsets of G) converge to which points $g \in G$. This convergence is written $\mathcal{F} \to g$.

DEFINITION: $\to$ is an *ℓ-convergence structure* on G if:

(a) $\dot{g} = \{X \subseteq G \mid g \in X\} \to g$ for all $g \in G$,

(b) $\mathcal{F} \supseteq \mathcal{M} \to g$ implies $\mathcal{F} \to g$,

(c) $\mathcal{F}, \mathcal{M} \to g$ implies $\mathcal{F} \cap \mathcal{M} \to g$,

(d) $\mathcal{F} \to g$, $\mathcal{M} \to h$ imply $\mathcal{F}\mathcal{M} \to gh$, $\mathcal{F} \vee \mathcal{M} \to g \vee h$, $\mathcal{F} \wedge \mathcal{M} \to g \wedge h$, and $\mathcal{F}^{-1} \to g^{-1}$.

We say that $(G,\to)$ is an *ℓ-convergence group* (ℓc-group). The general theory of convergence and Cauchy structures is undergoing development by Kent ([6], [7]) and others. Unless otherwise specified, all results of this section have proofs which may be found in [1] or [2].

An ℓ-convergence structure is determined by the filters which converge to 1.

PROPOSITION 2.1. Suppose $(G,\to)$ is an ℓc-group and that $\mathcal{V}$ is the set of filters convergent to 1. Then

(a) $\dot{1} \in \mathcal{V}$,

(b) $\mathcal{F} \supseteq \mathcal{M} \in \mathcal{V}$ implies $\mathcal{F} \in \mathcal{V}$,

(c) $\mathcal{F}, \mathcal{M} \in \mathcal{V}$ implies $\mathcal{F} \cap \mathcal{M} \in \mathcal{V}$,

(d) $\mathcal{F} \in \mathcal{V}$ implies $\mathcal{F}^2$, $\mathcal{F}^{-1} \in \mathcal{V}$,

(e) $\mathcal{F} \in \mathcal{V}$ implies $g^{-1}\mathcal{F}g \in \mathcal{V}$ for all $g \in G$,

(f) $\mathcal{F} \in \mathcal{V}$ and $a \wedge b = 1$ imply $\mathcal{F}a \wedge \mathcal{F}b \in \mathcal{V}$.

Conversely, if $\mathcal{V}$ is a collection of filters satisfying these conditions then the unique ℓ-convergence structure inducing $\mathcal{V}$ is obtained by declaring $\mathcal{F} \to g$ if $\mathcal{F}g^{-1} \in \mathcal{V}$.

Here are three natural and useful ℓ-convergence structures, each defined on an arbitrary ℓ-group G:

$\mathcal{F}$ *order converges* to 1, written $\mathcal{F} \overset{o}{\to} 1$, if $\bigwedge\{t \mid t^{-1} \leq F \leq t \text{ some } F \in \mathcal{F}\} = 1$.

$\mathcal{F}$ *polar converges* to 1, written $\mathcal{F} \overset{p}{\to} 1$, if $\cap\{F^{\perp\perp} \mid F \in \mathcal{F}\} = 1$.

$\mathcal{F}$ *α-converges* to 1, written $\mathcal{F} \overset{\alpha}{\to} 1$, when $[\bigvee(F \wedge x) = \bigwedge(F \vee x) = x$ for all $F \in \mathcal{F}]$ iff $x = 1$.

An ℓ-convergence structure $\to$ is *Hausdorff* if $\mathcal{F} \to 1$ implies $|\cap\mathcal{F}| \leq 1$; $\to$ is *convex* if $\mathcal{F} \to 1$ implies $\mathcal{F}^{\sim} = \{Y \mid Y \supseteq F^{\sim}, F \in \mathcal{F}\} \to 1$ (here $F^{\sim} = \{g \mid f_1 \leq g \leq f_2, f_i \in F\}$); $\to$ is *order closed* if $\mathcal{F} \to 1$ implies ocl $(\mathcal{F}) = \{Y \mid Y \supseteq \text{ocl}(F), F \in \mathcal{F}\} \to 1$; $\to$ is *strongly normal* if $\mathcal{F}^{-1}\mathcal{F}, \mathcal{F}\mathcal{F}^{-1}, \mathcal{H} \to 1$ imply $\mathcal{F}^{-1}\mathcal{H}\mathcal{F} \to 1$. All three above mentioned ℓ-convergence structures have these properties.

Suppose $\overset{x}{\to}$ is an ℓ-convergence structure. Define a filter $\mathcal{F}$ to be *Cauchy* if $\mathcal{F}\mathcal{F}^{-1}, \mathcal{F}^{-1}\mathcal{F} \overset{x}{\to} 1$. For Cauchy filters $\mathcal{F}$ and $\mathcal{M}$ define $\mathcal{F} \sim \mathcal{M}$ if $\mathcal{F} \cap \mathcal{M}$ is Cauchy. For Cauchy $\mathcal{F}$, let $[\mathcal{F}] = \{\mathcal{M} \mid \mathcal{M} \sim \mathcal{F}\}$ and let $G^x = \{[\mathcal{F}] \mid \mathcal{F}$ is Cauchy$\}$. For $[\mathcal{F}], [\mathcal{M}] \in G^x$ define $[\mathcal{F}][\mathcal{M}] = [\mathcal{F}\mathcal{M}]$, $[\mathcal{F}] \vee [\mathcal{M}] = [\mathcal{F} \vee \mathcal{M}]$, $[\mathcal{F}] \wedge [\mathcal{M}] = [\mathcal{F} \wedge \mathcal{M}]$, and $[\mathcal{F}]^{-1} = [\mathcal{F}^{-1}]$.

THEOREM 2.2. If $\overset{x}{\to}$ is a convex strongly normal ℓ-convergence structure, then G^x is an ℓ-group. If $\overset{x}{\to}$ is also Hausdorff, then the map dot: $G \to G^x$ defined by $(g)\text{dot} = [\dot{g}]$ is an ℓ-monomorphism. If in addition $\overset{x}{\to}$ is order closed then G is order dense in G^x.

So far none of our requirements would prevent the consideration of an ℓ-convergence structure whose behavior might be grossly inconsistent from group to group. Any satisfactory completion theory requires a modicum of consistency; the following three properties, all of which are enjoyed by $\overset{o}{\to}$, $\overset{p}{\to}$, and $\overset{\alpha}{\to}$, are minimal.

C_1: If $\theta: G \to H$ is an ℓ-isomorphism then $\mathcal{F} \overset{x}{\to} 1$ implies $\mathcal{F}\theta \overset{x}{\to} 1$.

C_2: If G is large in H then the restriction to G of $\overset{x}{\to}$ on H is coarser than $\overset{x}{\to}$ on G.

C_3: If $h \in G^x$ and $\mathcal{F}$ is a filter on G then $\mathcal{F} \overset{x}{\to} h$ iff $h = [\mathcal{F}]$.

An ℓ-convergence structure which is convex, Hausdorff, strongly abelian, order closed, and which satisfies C_1, C_2, and C_3 will be termed a *proper convergence*.

If $\overset{x}{\to}$ is any ℓ-convergence structure on H and if $G \subseteq H$ then the *closure* of G in H is $\{h \in H | G \in \mathcal{F} \to h\}$, written $cl_H(G)$. G is *closed* if $G = cl_H(G)$. Note that $cl_H(G)$ is not necessarily closed--the smallest closed set containing G is called the *iterated closure* of G in H, written $itcl_H(G)$. The iterated closure may also be obtained by iterating the closure operator, taking unions at limit stages.

THEOREM 2.3. Suppose $\overset{x}{\to}$ is a proper convergence. If $\psi: G \to H$ is an ℓ-monomorphism such that $G\psi$ is large in H then there is a unique ℓ-isomorphism $\psi^\wedge$ mapping G^x onto $cl(G\psi)$ in H^x such that $\psi^\wedge$ extends ψ.

Proof. Consider $f \in G^x$. By definition $f = [\mathcal{F}]$ for some Cauchy filter $\mathcal{F}$ on G. By C_1 and C_2 $\mathcal{F}\psi$ is Cauchy on H. Define $(f)\psi^\wedge = [\mathcal{F}\psi]$, clearly an ℓ-monomorphism. $\psi^\wedge$ is one-to-one by proposition 2.9 of [1]. By C_3, $\mathcal{F}\psi \overset{x}{\to} [\mathcal{F}\psi] \in cl(G\psi)$. On the other hand, if $h \in H^x$ then there is some filter $\mathcal{H}$ on H such that $h = [\mathcal{H}]$. By C_1, $\mathcal{H}\psi^{-1}$ is Cauchy on G so $[\mathcal{H}\psi^{-1}] \in G^x$. Since $[\mathcal{H}\psi^{-1}]\psi^\wedge = [\mathcal{H}] = h$, $\psi^\wedge$ maps G^x onto $cl(G\psi)$ in H^x.

Suppose $\overset{x}{\to}$ is a proper convergence. An ℓ-group G is *x-complete* if $G^x = G$. H is an *x-completion* of G if G is large in H, if H is x-complete, and if $G \leq K < H$ implies K is not x-complete.

THEOREM 2.4. Suppose $\overset{x}{\to}$ is a proper convergence. An ℓ-group G has an x-completion iff G is large in some x-complete ℓ-group H, in which case the x-completion is $itcl_H(G)$.

Proof. Let L be $itcl_H(G)$. By theorem 2.3, $L^x \leq cl_H(L) = L$, so L is x-complete. If $G \leq M < L$ then $cl_H(M) \neq M$. If ψ is the identity map on M then the fact that the extension $\psi^\wedge: M^x \to cl_H(M)$ is onto implies that M is not x-complete.

THEOREM 2.5. Suppose $\overset{x}{\to}$ is a proper convergence. Then H is an x-completion of G iff G is large in H, H is x-complete, and any ℓ-monomorphism ψ from G onto a large ℓ-subgroup of the x-complete ℓ-group M can be uniquely extended to an ℓ-monomorphism $\psi^\wedge: H \to M$.

Proof. Suppose H is an x-completion of G. Theorem 2.4 says $itcl_H(G) = H$. Let $G = G_0 \leq G_1 \leq G_2 \ldots \leq G_\alpha \leq \ldots \leq G_\gamma = H$ be a sequence of ℓ-subgroups of H such that $G_{\alpha+1} = cl_H(G_\alpha)$ and $G_\alpha = \bigcup \{G_\beta | \beta < \alpha\}$ for limit ordinals α. Given $\psi: G \to M$, repeated use of Theorem 2.4 produces a unique ℓ-monomorphism $\psi^\wedge: H \to M$. The converse is clear.

COROLLARY 2.6. Suppose $\overset{x}{\to}$ is a proper convergence. If G has an x-completion then it is unique up to an ℓ-isomorphism over G.

If G has an x-completion, we designate it G^{ix}, for iterated x.

Now we are up against a most interesting open question. Is G^x x-complete? That is, is $G^{xx} = G^x$? Nothing we have said so far seems to guarantee it, though no one has succeeded in producing an ℓ-group for which $G^{xx} \neq G^x$, where x is an ℓ-convergence structure which is convex, Hausdorff, strongly normal, and order closed. Positive results in special cases are known; for example, $G^{oo} = G^o$ for every ℓ-group, and $G^{\alpha\alpha} = G^\alpha$ for completely distributive ℓ-groups. It is also known that G^{xx} need not equal G^x if the hypothesis of order closure of $\overset{x}{\to}$ is dropped.

One approach to difficult problems is to avoid them. Instead of settling whether $G^{xx} = G^x$, let us iterate the process freely, hoping it will eventually stabilize. The necessary tool for this purpose is the concept of $\preccurlyeq$ extension. Define $G \preccurlyeq H$ to mean that G is order dense in H and that $\bigwedge\{|hg^{-1}| \,|\, g \in G\} = 1$ for each $h \in H$.

Proposition 2.9 and half of proposition 2.7 are quite easy to verify; proposition 2.10 and the remainder of 2.7 are quite difficult. Proofs can be found in [2].

PROPOSITION 2.7. Suppose $G \leq H \leq K$. Then $G \preccurlyeq K$ iff $G \preccurlyeq H \preccurlyeq K$.

PROPOSITION 2.8. If $\overset{x}{\to}$ is a proper convergence, then $G \preccurlyeq G^x$.

Proof. By corollary 3.7, $G^x \leq G^\alpha$, so that by the previous proposition it is enough to show $G \preccurlyeq G^\alpha$. Given $h \in G^\alpha$ and $1 < d \in G$ we seek $g \in G$ such that $|hg^{-1}| \not\geq d$. Now, $h = [\mathcal{F}]$, where $\mathcal{F}$ is a filter on G Cauchy with respect to $\overset{\alpha}{\to}$. By C_3, $\mathcal{F} \overset{\alpha}{\to} h$, implying $h\mathcal{F}^{-1} \vee \mathcal{F}h^{-1} \overset{\alpha}{\to} 1$. It follows from [2] that there is some $F \in \mathcal{F}$ and t such that $(hF^{-1} \vee Fh^{-1}) \wedge d \leq t < d$. Choosing $g \in F$, we get $|hg^{-1}| \wedge d \leq t < d$, so $|hg^{-1}| \not\geq d$.

PROPOSITION 2.9. Suppose $\mathcal{C}$ is a collection of ℓ-groups totally ordered by $\preccurlyeq$. Then $C \preccurlyeq \bigcup\mathcal{C}$ for any $C \in \mathcal{C}$.

PROPOSITION 2.10. If $G \preccurlyeq H$ then $|H| \leq |2^G|$.

PROPOSITION 2.11. Suppose $\overset{x}{\to}$ is a proper convergence. Then every ℓ-group G has an x-completion G^{ix} which is unique up to an ℓ-isomorphism over G.

Proof. Given G let γ be an ordinal of cardinality greater than $|2^G|$. Define $G_0 = G$, $G_{\alpha+1} = G_\alpha{}^x$ and $G_\delta = \bigcup\{G_\alpha | \alpha < \delta\}$ for limit ordinals δ. By propositions 2.8 and 2.9, $G \preccurlyeq G_\alpha$ for all α. By proposition 2.10, $G_\gamma = G_{\gamma+1} = G^{ix}$.

So proper convergences provide usable Cauchy completions. It is time now to look at the cut completeness properties of these completions.

3. SOME CUT COMPLETIONS

The most tractable of the three mentioned convergence structures is the order convergence structure. Not only is $G^o = G^{oo} = G^{io}$, but in this case the Cauchy completion G^o exactly coincides with the flip flop cut completion. In the next theorem the notation $G \leq_o H$ means $G \leq H$ and $\bigvee L_G(h) = h$ for all $h \in H$.

THEOREM 3.1. For $G \leq H$ the following are equivalent.

(a) H is ℓ-isomorphic to G^o over G,

(b) H is a maximal $\leq_o$ extension of G,

(c) G is large in H, H is flip flop cut complete, and $G \leq K < H$ implies K is not flip flop cut complete,

(d) H is ℓ-isomorphic to the ℓ-group of flip flop cuts of G,

(e) every complete ℓ-homomorphism $\psi: G \to M$ can be uniquely extended to a complete ℓ-homomorphism $\psi^\wedge: H \to M^o$.

Proof. Most of the details of this proof may be found in [1]. We show only how a supremum for each flip flop cut X of G is produced in G^o. For each $x \in X$ and $u \in U(X)$ define the *interval* $[x,u]$ to be $\{g \in G | x \leq g \leq u\}$. Let $\mathcal{F}$ be the filter generated by such intervals. Since $xu^{-1} \leq [x,u]\cdot[x,u]^{-1}$ and $[x,u]\cdot[x,u]^{-1} \leq ux^{-1}$, and since $\bigwedge U(X)\cdot X^{-1} = 1$, $\mathcal{F}\mathcal{F}^{-1} \overset{o}{\to} 1$. Similarly, $\mathcal{F}^{-1}\mathcal{F} \overset{o}{\to} 1$. Therefore $[\mathcal{F}] \in G^o$, and it is easily shown that $[\mathcal{F}] = \bigvee X$ in G^o.

G^o has been extensively studied in the archimedean case [5] where its completeness properties may be put more strongly. An archimedean ℓ-group G is flip flop cut complete if it is *conditionally complete*: every subset of H which has an upper bound has a least upper bound. In this case G^o is called the *Dedekind MacNeille* completion of G.

The polar convergence structure is more problemmatical. Whatever else is true, however, G^{ip} does have the property of locally finite cut completeness.

PROPOSITION 3.2. If X is a locally finite cut of G then X has a supremum in G^p.

Proof. Though a general proof is available in [1], the main idea can be understood by seeing how a supremum for a disjoint subset D of G is "manufactured" in G^p. For each finite $A \subseteq D$ let $F_A = \{\bigvee B \mid B \text{ finite}, A \subseteq B \subseteq D\}$ and let $\mathcal{F}$ be the filter generated by the F_A's. Because D is a disjoint set, $F_A F_A^{-1} \subseteq a^{\perp}$ for each $a \in A$, which implies $\bigcap (F_A F_A^{-1})^{\perp\perp} = 1$. That is, $\mathcal{F}$ is Cauchy. In fact, one can easily show that $[\mathcal{F}] = \bigvee D$ in G^p.

THEOREM 3.3. Every ℓ-group G has a locally finite cut completion which is unique up to an ℓ-isomorphism over G.

Proof. Let $U = \{K \mid G \leq K \leq G^{ip}$ and K is locally finite cut complete$\}$. Clearly $G^{ip} \in U$ and $\bigcap U \in U$. In fact, $\bigcap U$ is the completion we seek.

Only a little seems to be presently understood about G^p or G^{ip}.

THEOREM 3.4. Suppose G is archimedean or strongly projectable. For $G \leq H$ the following are equivalent:

(a) H is ℓ-isomorphic to G^p over G,

(b) H is ℓ-isomorphic to G^{ip} over G,

(c) H is the orthocompletion of G,

(d) H is the lateral completion of G,

(e) H is the locally finite cut completion of G.

The lack of any idea of how to prove a conjecture should never deter one from posing it. So why shouldn't G^{ip} be the locally finite cut completion in general? The coincidence of these two completions is a truly pleasing possibility.

The most interesting Cauchy completion of all is G^{α}. For one thing, G^{α} is the largest of the Cauchy completions G^{x} with respect to proper convergences $\overset{x}{\rightarrow}$. That is because of the striking fact that $\overset{\alpha}{\rightarrow}$ is the coarsest convex Hausdorff order closed ℓ-convergence structure on any ℓ-group G.

THEOREM 3.5. Suppose $\overset{x}{\rightarrow}$ is a convex Hausdorff order closed ℓ-convergence structure and $\mathcal{F}$ is a filter on G. Then $\mathcal{F} \overset{x}{\rightarrow} 1$ implies $\mathcal{F} \overset{\alpha}{\rightarrow} 1$.

If G is completely distributive then the hypothesis that $\overset{x}{\rightarrow}$ be order closed can be dropped in Theorem 3.5. In this form the result is due to Madell [8]. A proof of Theorem 3.5 as stated can be found in [2].

PROPOSITION 3.6. Suppose $\overset{x}{\rightarrow}$ is a convex Hausdorff strongly normal order closed ℓ-convergence structure. Then the map $\theta : G^{x} \rightarrow G^{\alpha}$ defined by $[\mathcal{F}]_x \theta = [\mathcal{F}]_{\alpha}$ is an ℓ-monomorphism.

Proof. The crucial one-to-oneness of θ is a direct result of the order closedness of $\overset{x}{\rightarrow}$ by way of Proposition 2.9 of [1].

COROLLARY 3.7. $G^{x} \leq G^{\alpha}$ for every proper convergence $\overset{x}{\rightarrow}$.

COROLLARY 3.8. An ℓ-group G is α-complete iff it is x-complete for every proper convergence x.

Such general Cauchy completeness must connote correspondingly general cut completeness.

THEOREM 3.9. Every cut X of G has a supremum in G^{α}.

Proof. For each $x \in X$ let $F_x = \{y \in X | y \geq x\}$ and let $\mathcal{F}$ be the filter generated by the F_x's. In order to show $\mathcal{F}$ Cauchy consider $g > 1$. Since $gX \neq X = X'$ there must be some $x \in X$ and y such that $gx \wedge x_1 \leq y < gx$ for all $x_1 \in X$. Therefore $g \wedge x_1 x_2^{-1} \leq yx^{-1} < g$ for all $x_1, x_2 \in F_x$. This shows $\mathcal{F}\mathcal{F}^{-1} \overset{\alpha}{\rightarrow} 1$ and the argument for $\mathcal{F}^{-1}\mathcal{F} \overset{\alpha}{\rightarrow} 1$ is analogous. It is then routine to verify that $[\mathcal{F}] = \bigvee X$ in G^{α}.

THEOREM 3.10. Given any ℓ-group G there is an ℓ-group G^{s} unique up to an ℓ-isomorphism over G, having the following equivalent properties.

(a) G^s is the cut completion of G,

(b) If $G \leqq K$ then there is a unique complete ℓ-monomorphism $\psi: ocl_K(G) \to G^s$ such that ψ is the identity on G. In addition, if $G \leq L < G^s$ then L fails to have this property,

(c) G is large in G^s, $ocl(G) = G^s$, and G^s is order closed in every ℓ-group in which it is completely embedded.

Proof. G^s is $\bigcap\{A \mid G \leq A \leq G^{i\alpha}$ and A is cut complete$\}$.

The conjecture that goes with this result is that $G^{i\alpha}$ is the cut completion; that is, that $ocl(G) = G^{i\alpha}$. This result would be the most natural generalization of the coincidence of Cauchy and cut completions of the rational numbers.

Now that we know the cut completion exists, an obvious construction suggests itself. G is an *ℓ-monoid* if G is a semigroup with identity and a lattice such that $a \leq b$ implies $ag \leq bg$ and $ga \leq gb$ for $a,b,g \in G$. $X \subseteq G$ is a *lower cut (upper cut)* of G if X is an order closed element ideal (element filter) such that $Xc \neq X$ and $cX \neq X$ for any $1 < c \in G$. For lower cuts X_1 and X_2 define $X_1 * X_2$ to be $(X_1X_2)'$ and define $X_1 \leq X_2$ to mean $X_1 \subseteq X_2$. Define analogous operations on upper cuts. Let LC(G) (UC(G)) be the set of lower cuts (upper cuts) of G. Let A(G) and B(G) be the invertible elements of UC(LC(G)) and LC(UC(G)) respectively.

LEMMA 3.11. Suppose G is an ℓ-group. Then LC(G) and UC(G) are ℓ-monoids containing G. A(G) and B(G) are ℓ-groups containing respectively LC(G) and UC(G). Finally $G \preccurlyeq A(G)$ and $G \preccurlyeq B(G)$.

By imagining the constructions to be taking place inside G^s, one sees the preceding lemma to be true. A direct construction of G^s, however, would require a direct proof, which is surprisingly difficult.

Given an ℓ-group G, define $G_0 = G$, $G_{\alpha+1} = A(G)$ for α even, $G_{\alpha+1} = B(G)$ for α odd, and $G_\delta = \bigcup\{G_\alpha \mid \alpha < \delta\}$ for limit ordinals δ. By propositions 2.7, 2.9, and 2.10, whose proofs do not mention Cauchy or convergence spaces, $G_\delta = G_{\delta+1} = G_{\delta+2}$ for any ordinal δ of cardinality greater than $|2^G|$. Thus we have a direct construction of $G^s = G_\delta$.

4. QUESTIONS NOT ALREADY ASKED

1. The Cauchy constructions outlined here and given in detail in [1] have an ad hoc flavor. Is this the correct way to do Cauchy constructions? This question is the subject of [6].

2. What are the structures of the various completions described here? Such results would presumably make use of the classical ℓ-group representation theorems.
3. Many questions can be asked about $\overset{\alpha}{\rightarrow}$, $\overset{\ell}{\rightarrow}$, and $\overset{p}{\rightarrow}$. For example, when is $\overset{\ell}{\rightarrow}$ topological? When are any of the above pseudotopological?
4. There are many natural questions concerning the relationships between the various Cauchy completions. For example, is $(G^{\alpha})^{ip} = G^{i\alpha}$?

REFERENCES

1. R. Ball, Convergence and Cauchy structures on lattice ordered groups. *Trans. Amer. Math. Soc. (2) 259:* 357-392 (1980).

2. R. Ball and G. Davis, The α-completion of a lattice ordered group. In press.

3. S. Bernau, The lateral completion of an arbitrary lattice group. *J. Australian Math. Soc. 19A:* 263-289 (1975).

4. P. Conrad, The lateral completion of a lattice-ordered group. *Proc. London Math. Soc. (3) 19:* 444-480 (1969).

5. P. Conrad and D. McAlister, The completion of a lattice ordered group. *J. Australian Math. Soc. 9:* 182-208 (1969).

6. D. Kent, Completions of ℓ-Cauchy groups, this volume. *of the Boise State Conference*, Marcel Dekker.

7. D. Kent and G. Richardson, Regular completions of Cauchy spaces. *Pacific J. Math. 51:* 483-490 (1974).

8. R. Madell, Complete distributivity and α-convergence. Unpublished, Village Community School, 272 West Tenth Street, New York, N.Y. 10014.

COMPLETIONS OF ℓ-CAUCHY GROUPS

D. C. Kent
Washington State University
Pullman, Washington

R. N. Ball has obtained significant results on completions of ℓ-groups by employing Cauchy structures derived from certain convergence structures intrinsic to the algebraic structure of the ℓ-group. His constructions are not Cauchy space completions in the ordinary sense because they do not involve the construction of a complete Cauchy structure on the completion space. However Ball's results certainly suggest that the development of a general Cauchy space completion theory for ℓ-groups would be a worthwhile undertaking, and this note represents a step in this direction.

Background information on Cauchy spaces, convergence spaces, and ℓ-groups will not be reviewed here; the reader is referred to [1] and [2] for additional information on the first two topics, and to [1] for information on the third.

Let $(G, \cdot , \vee , \wedge)$ be an ℓ-group (i.e., a lattice-ordered group); hereafter, we shall refer merely to the ℓ-group G. Let F(G) be the set of all filters on G, and let $\dot{x}$ be the fixed ultrafilter generated by x. Consider the following axioms for a subset $\mathcal{C}$ of F(G).

(C_1) For each x in G, $\dot{x} \in \mathcal{C}$.

(C_2) If $\mathcal{F} \in \mathcal{C}$, $\mathcal{G} \in F(G)$, and $\mathcal{F} \subset \mathcal{G}$, then $\mathcal{G} \in \mathcal{C}$.

(C_3) If $\mathcal{F}, \mathcal{G} \in \mathcal{C}$, then $\mathcal{F}^{-1}\mathcal{G}$ and $\mathcal{F}\mathcal{G}^{-1} \in \mathcal{C}$.

(C_4) If $\mathcal{F}, \mathcal{G} \in \mathcal{C}$, then $\mathcal{F} \vee \mathcal{G} \in \mathcal{C}$ and $\mathcal{F} \wedge \mathcal{G} \in \mathcal{C}$.

(C_5) If $\dot{x}, \dot{y} \in \mathcal{C}$, then x = y.

In [1], $\mathcal{C}$ is defined to be a *group Cauchy structure* on G if conditions (C_1), (C_2), and (C_3) are satisfied, and a group Cauchy structure which satisfies (C_4) is called an *ℓ-Cauchy structure*. A group Cauchy structure which satisfies (C_5) is said to be T_2. We define an *ℓ-Cauchy group* $(G,\mathcal{C})$ to be an ℓ-group G equipped with a T_2 ℓ-Cauchy structure $\mathcal{C}$. A *Cauchy structure* can be defined on any set by imposing axioms (C_1) and (C_2) along with condition:

$(C_3)'$ If $\mathcal{F}$, $\mathcal{G} \in \mathcal{C}$ and $\mathcal{F}$ and $\mathcal{G}$ fail to contain disjoint sets, then $\mathcal{F} \cap \mathcal{G} \in \mathcal{C}$. A group Cauchy structure is a special case of a Cauchy structure.

If $\mathcal{C}$ is a group Cauchy structure on G, then two Cauchy filters $\mathcal{F}$, $\mathcal{G} \in \mathcal{C}$ are said to be *equivalent* (written $\mathcal{F} \sim \mathcal{G}$) if $\mathcal{F} \cap \mathcal{G} \in \mathcal{C}$, or, equivalently, $\mathcal{F}^{-1}\mathcal{G} \in \mathcal{C}$. If $\mathcal{F} \sim \dot{x}$, then we write "$\mathcal{F} \to x$", meaning that $\mathcal{F}$ converges to x in the convergence structure on G associated with $\mathcal{C}$. A Cauchy structure is *complete* if every Cauchy filter is convergent. A *completion* $((\hat{G},\hat{\mathcal{C}}), g)$ of an ℓ-Cauchy group $(G,\mathcal{C})$ consists of a complete ℓ-Cauchy group $(\hat{G},\hat{\mathcal{C}})$ and a map $g: G \to \hat{G}$ which is both a Cauchy embedding and an ℓ-group isomorphism onto an ℓ-subgroup of $\hat{G}$ such that $g(G)$ is dense in $\hat{G}$. The latter term means that every point in $\hat{G}$ is the limit of a filter containing $g(G)$.

Let $(G,\mathcal{C})$ be an ℓ-Cauchy group. If $\mathcal{F} \in \mathcal{C}$, let $[\mathcal{F}] = \{\mathcal{G} \in \mathcal{C}: \mathcal{F} \sim \mathcal{G}\}$, and let $G^* = \{[\mathcal{F}]: \mathcal{F} \in \mathcal{C}\}$. The natural map $j: G \to G^*$ is defined by $j(x) = [\dot{x}]$. Define ℓ-group operations on G^* as follows: $[\mathcal{F}][\mathcal{G}] = [\mathcal{F}\mathcal{G}]$, $[\mathcal{F}]^{-1} = [\mathcal{F}^{-1}]$, $[\mathcal{F}] \vee [\mathcal{G}] = [\mathcal{F} \vee \mathcal{G}]$, and $[\mathcal{F}] \wedge [\mathcal{G}] = [\mathcal{F} \wedge \mathcal{G}]$. The following proposition is proved in Section 2 of [1].

PROPOSITION 1. For any ℓ-Cauchy group $(G,\mathcal{C})$, G^* is an ℓ-group and $j: G \to G^*$ is an ℓ-isomorphism onto an ℓ-subgroup of G^*.

We shall construct an ℓ-Cauchy structure $\mathcal{C}^*$ on G^*. For convenience, let $a_{\mathcal{F}} = [\mathcal{F}]$ denote the member of G^* determined by a filter $\mathcal{F} \in \mathcal{C}$. If $\mathcal{F} \to x$, then $a_{\mathcal{F}} = j(x) = [\dot{x}]$. For each free filter $\mathcal{F} \in \mathcal{C}$ (a free filter is one in which no point belongs to every set in the filter), let $\Phi_{\mathcal{F}} = j(\mathcal{F}) \cap \dot{a}_{\mathcal{F}} \in F(G^*)$. A filter Φ in $F(G^*)$ will be called *subbasic for* $\mathcal{C}^*$ if $\Phi = \Phi_{\mathcal{F}}$ for some free filter $\mathcal{F} \in \mathcal{C}$, or else Φ is a fixed ultrafilter on G^*. Note that inverses of subbasic filters are again subbasic. Any finite product of subbasic filters will be called a *prebasic filter for* $\mathcal{C}^*$. Let β^* be the smallest subset of $F(G^*)$ closed under the lattice operations on G^* which contains all prebasic filters; members of β^* are called *basic filters for* $\mathcal{C}^*$. Finally, we define $\mathcal{C}^* = \{\Phi \in F(G^*): \psi \subset \Phi \text{ for some } \psi \in \beta^*\}$.

THEOREM 2. For any ℓ-Cauchy group $(G,\mathcal{C})$, $((G^*,\mathcal{C}^*), j)$ is an ℓ-Cauchy group completion of $(G,\mathcal{C})$.

Proof. It is obvious that $\mathcal{C}^*$ satisfies conditions (G_1), (G_2), and (G_4). The basic elements of $\mathcal{C}^*$ are obtained by starting with prebasic filters and successively applying the lattice operations. Since multiplication is distributive relative to the lattice operations, it follows that the basic filters are closed under the group operations; thus $\mathcal{C}^*$ satisfies (G_3).

We next make the observation that every prebasic filter Φ is either a fixed ultrafilter or else a filter of the form $\psi \cap \dot{b}$, where ψ is a free filter on G^* and $\dot{b} \in G^*$. This follows from the fact that Φ is a product of n subbasic filters, each of which is contained in exactly one fixed ultrafilter $\dot{a}_k$; $b = a_1 a_2 \ldots a_n$. Note that if $\Phi = \psi \cap \dot{b}$, then $\Phi \to b$. If $\{\Phi_k\}$ is a finite set of prebasic filters, each converging to b_k, then $\vee \{\Phi_k\} \to \vee b_k$ and $\wedge \{\Phi_k\} \to \wedge b_k$. This reasoning can be extended to show that $(G^*, \mathcal{C}^*)$ is complete. It also follows from the above characterization of prebasic filters that the set of prebasic filters satisfies (G_5), and it is easy to see that this property extends to $\mathcal{C}^*$ as well.

If $a_{\mathcal{F}} \in G^* - j(G)$, then $j(\mathcal{F}) \cap \dot{a}_{\mathcal{F}}$ is a subbasic member of $\mathcal{C}^*$, and so $j(\mathcal{F}) \to a_{\mathcal{F}}$; thus $j(G)$ is dense in G^*.

If $\mathcal{F} \in \mathcal{C}$, then $j(\mathcal{F})$ contains a subbasic filter, and is therefore in $\mathcal{C}^*$; thus j is Cauchy-continuous. To show that j^{-1} is Cauchy-continuous, first recall that the subbasic members of $\mathcal{C}^*$ are filters of the form $j(\mathcal{F})$ for $\mathcal{F} \in \mathcal{C}$, adjoined to fixed ultrafilters on G^*. Any filter Φ in $\mathcal{C}^*$ which contains $j(G)$ is obtained by applying the ℓ-group operations to subbasic filters. Since $j^{-1} : j(G) \to G$ is an ℓ-isomorphism, it follows that $j^{-1}(\Phi) \in \mathcal{C}$, and the proof is complete.

THEOREM 3. Let $(G, \mathcal{C})$ be an ℓ-Cauchy group, let $(G_1, \mathcal{C}_1)$ be a complete ℓ-Cauchy group, and let $f:(G,\mathcal{C}) \to (G_1, \mathcal{C}_1)$ be a Cauchy-continuous ℓ-homomorphism. Then there is a unique Cauchy-continuous ℓ-homomorphism f^* which makes the following diagram commute:

$$\begin{array}{ccc} (G,\mathcal{C}) & \xrightarrow{j} & (G^*,\mathcal{C}^*) \\ & \searrow f & \downarrow f^* \\ & & (G_1,\mathcal{C}_1) \end{array}$$

Proof. For $a_{\mathcal{F}} \in G^*$, define $f^*(a_{\mathcal{F}}) = y$ if and only if $f(\mathcal{F}) \to y$ in G_1. It is a simple matter to verify that f^* is a well-defined map. To check that f^* is an ℓ-homomorphism, let $a_{\mathcal{F}}, a_{\mathcal{G}} \in G^*$, and let $f^*(a_{\mathcal{F}}) = x$, $f^*(a_{\mathcal{G}}) = y$. Then $f(\mathcal{F}) \cap \dot{x}$ and $f(\mathcal{G}) \cap \dot{y}$ are both in $\mathcal{C}_1$, which implies that $f(\mathcal{F}\mathcal{G}^{-1}) \cap (\dot{xy}^{-1}) \supset (f(\mathcal{F}) \cap \dot{x})(f(\mathcal{G}) \cap \dot{y})^{-1} \in \mathcal{C}^{-1}$. Therefore $f^*(a_{\mathcal{F}})f^*(a_{\mathcal{G}}-1) = xy^{-1} =$

$f^*(a_{\mathscr{F}\mathscr{G}^{-1}}) = f^*(a_{\mathscr{F}}\, a_{\mathscr{G}^{-1}})$, and f^* preserves the group operations. Similar arguments show that f^* preserves the lattice operations.

To show that f^* is Cauchy-continuous, first note that $f^*(\Phi) \in \mathscr{C}_1$, for any subbasic filter $\Phi \in \mathscr{C}^*$. But any other filter $\psi \in \mathscr{C}^*$ is obtained by applying the ℓ-group operations to the subbasic filters, and $f^*(\psi) \in \mathscr{C}_1$ follows from the fact that f^* is an ℓ-homomorphism.

The uniqueness of f^* results from f^* being uniquely determined on the dense ℓ-subgroup $j(G)$ of G^* and the fact that all spaces involved are T_2.

If $((\hat{G}, \hat{\mathscr{C}}), g)$ and $((\tilde{G}, \tilde{\mathscr{C}}), h)$ are ℓ-Cauchy completions of an ℓ-Cauchy group $(G, \mathscr{C})$, then the latter completion is said to be *larger* (or *finer*) than the former, written $((\hat{G}, \hat{\mathscr{C}}), g) \leq ((\tilde{G},\tilde{\mathscr{C}}), h)$, if there is a Cauchy-continuous ℓ-homomorphism k which makes the following diagram commute:

$$\begin{array}{ccc} (G,\mathscr{C}) & \xrightarrow{h} & (\tilde{G},\tilde{\mathscr{C}}) \\ & \searrow^{g} & \downarrow k \\ & & (\hat{G},\hat{\mathscr{C}}). \end{array}$$

If it happens that each of the completions is finer than the other, then they are said to be *equivalent*.

COROLLARY 4. For any ℓ-Cauchy group $(G,\mathscr{C})$, $((G^*,\mathscr{C}^*), j)$ is the largest (up to equivalence) ℓ-Cauchy group completion of $(G,\mathscr{C})$.

We conclude with a brief discussion of the role which Cauchy space completions might play in the further study of ℓ-groups.

Let LCH denote the category whose objects are ℓ-Cauchy groups and whose morphisms are Cauchy continuous ℓ-homomorphisms. Let LCH* be the full subcategory of LCH determined by the complete ℓ-Cauchy groups, and let T: LCH $\to$ LCH* be defined by $T(G,\mathscr{C}) = (G^*,\mathscr{C}^*)$ for each object $(G,\mathscr{C})$ in LCH, and for each morphism $f : (G_1,\mathscr{C}_1) \to (G_2,\mathscr{C}_2)$, let $T(f) = f^* : T(G_1,\mathscr{C}_1) \to T(G_2,\mathscr{C}_2)$ be the unique extension of f whose existence is guaranteed by Theorem 3. In the language of [2], T is called a *completion functor* on LCH.

A full subcategory MCH of LCH is called a *completion subcategory* of LCH if there is a completion functor defined on MCH. In other words, MCH is a completion subcategory if each ℓ-Cauchy group $(X,\mathscr{C})$ in MCH has an ℓ-group completion $((\hat{X},\hat{\mathscr{C}}), k)$ in MCH which is the largest completion relative to MCH.

The approach taken by Ball in [1] amounts to identifying certain subcategories of LCH whose members have Cauchy completions which are algebraically interesting. The next logical step in our approach would be to identify the completion subcategories of LCH, paying special attention to those which yield algebraically as well as topologically interesting completions. A top priority should be given to finding out whether the subcategories already studied by Ball are, or in some unique way determine, completion subcategories of LCH.

REFERENCES

1. Richard N. Ball, Convergence and Cauchy structures on lattice ordered groups. *Trans. Amer. Math. Soc. (2) 259:* 357-392 (1980).

2. R. Fric and D. C. Kent, Completion functors for Cauchy spaces. *Internatl. Math. J.* In press.

ℓ-GROUP VARIETIES

Jo E. Smith

Boise State University
*Boise, Idaho**

Let L be the collection of all ℓ-group varieties, where a variety of ℓ-groups is the collection of all ℓ-groups satisfying a given set of equations or equivalently is a collection of ℓ-groups closed with respect to ℓ-subgroups, cardinal products, and ℓ-homomorphic images. The collection L itself has a lattice ordering--set inclusion. Under this order, $\mathcal{U} \wedge \mathcal{V} = \mathcal{U} \cap \mathcal{V}$ for $\mathcal{U}, \mathcal{V} \in L$ and $\mathcal{U} \vee \mathcal{V} = \cap\{\mathcal{W} \in L | \mathcal{U} \cup \mathcal{V} \subseteq \mathcal{W}\}$. With such lattice operations L forms a dually Brouwer (and hence distributive) lattice [5], but not a Brouwerian lattice [7]. It is also possible to define an associative multiplication on L, where $\mathcal{U}\mathcal{V}$ is the collection of all ℓ-groups that contain an ℓ-ideal from $\mathcal{U}$ whose ℓ-quotient is in $\mathcal{V}$. This multiplication is compatible with the lattice operations of L. In fact, it has been shown that the multiplication distributes over all meets from either side, all joins from the right and finite joins from the left, and that it has cancellation properties [2].

Some of the better studied ℓ-group varieties are $\mathcal{E}$ the trivial variety, $\mathcal{A}$ the variety of abelian ℓ-groups, $\mathcal{W}$ the variety of weakly abelian ℓ-groups, $\mathcal{R}$ the variety of representable ℓ-groups, $\mathcal{N}$ the variety of normal-valued ℓ-groups, and $\mathcal{L}$ the variety of all ℓ-groups. These varieties are related as shown in the diagram of L on page 100. It has been shown that all nontrivial varieties contain $\mathcal{A}$ [8] and that all proper varieties are contained in $\mathcal{N}$ [3]. It is also known that the varieties $\mathcal{A}^2 = \mathcal{A}\mathcal{A}, \ldots, \mathcal{A}^n = \mathcal{A}^{n-1}\mathcal{A}$ form a proper chain from $\mathcal{A}$ to $\mathcal{N}$, with $\mathcal{N} = \bigvee\{\mathcal{A}^n | n \in N\}$ [2].

Since $\mathcal{A}$ plays such an important role in L, it is natural to attempt to generalize it. Therefore, for $n \geq 2$ let $\mathcal{L}_n$ be the variety defined by the

*Current address: Department of Science and Mathematics, General Motors Institute, Flint, Michigan

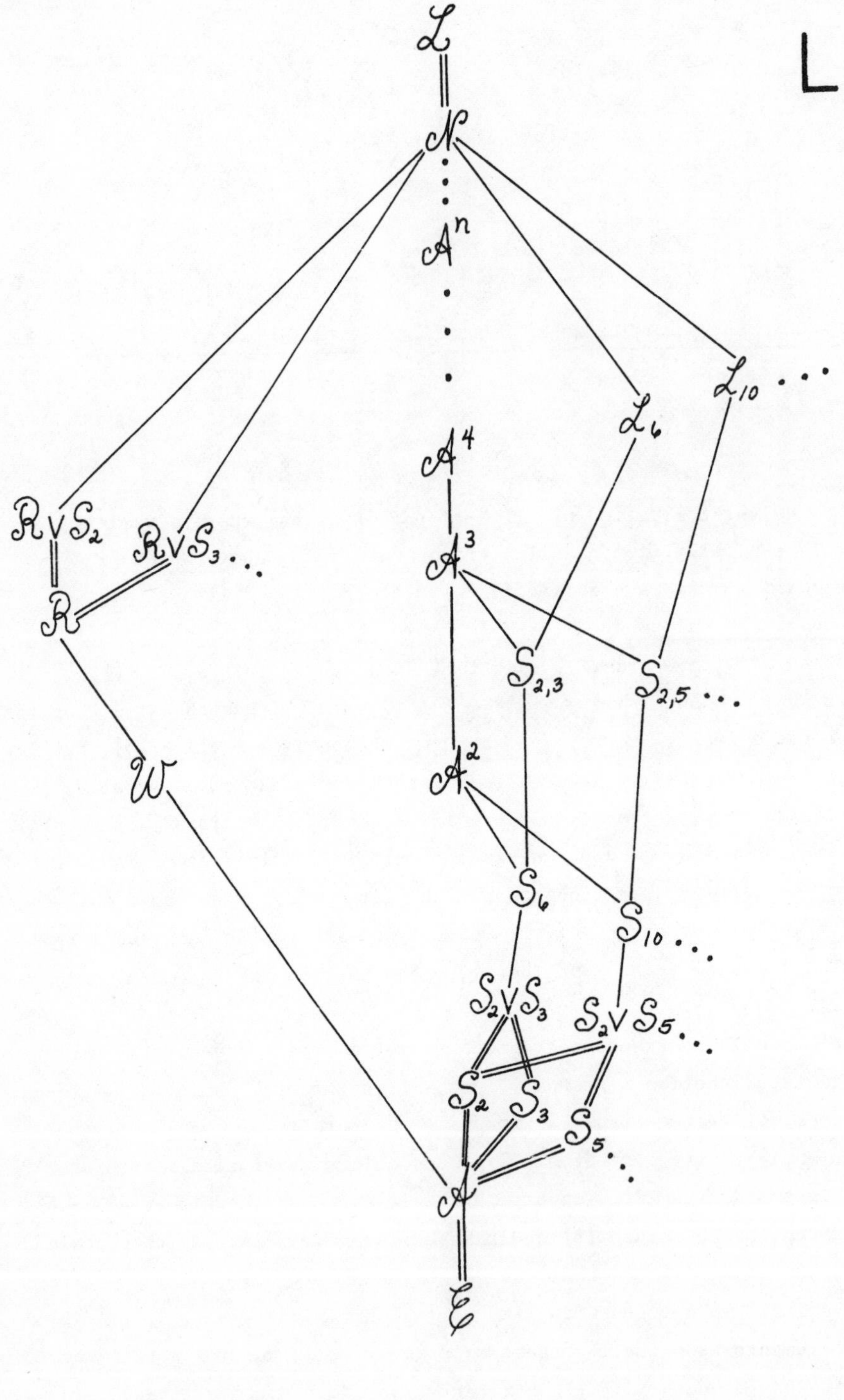
L
$\mathcal{L}$
$\mathcal{N}$
$\mathcal{A}^n$
$\mathcal{L}_6$
$\mathcal{L}_{10} \ldots$
$\mathcal{A}^4$
$\mathcal{R} \vee \mathcal{S}_2$
$\mathcal{R} \vee \mathcal{S}_3 \ldots$
$\mathcal{A}^3$
$\mathcal{R}$
$\mathcal{S}_{2,3}$
$\mathcal{S}_{2,5} \ldots$
$\mathcal{W}$
$\mathcal{A}^2$
$\mathcal{S}_6$
$\mathcal{S}_{10} \ldots$
$\mathcal{S}_2 \vee \mathcal{S}_3$
$\mathcal{S}_2 \vee \mathcal{S}_5 \ldots$
$\mathcal{S}_2$
$\mathcal{S}_3$
$\mathcal{S}_5 \ldots$
$\mathcal{A}$
$\mathcal{E}$

equation $[x^n, y^n] = 1$. These $\mathcal{L}_n$'s form a family with several properties of interest: $\mathcal{L}_m \subset \mathcal{L}_n$ iff m is a proper divisor of n [4], $\mathcal{L}_m \wedge \mathcal{L}_n = \mathcal{A}$ iff $(m,n) = 1$ [6], and $\mathcal{L}_n \wedge \mathcal{R} = \mathcal{A}$ for all $n \geq 2$ [4]. Also, within each $\mathcal{L}_n$ lies a subdirectly irreducible ℓ-group of particular interest:

$$G_n = \left(\prod_0^{n-1} \mathbb{Z}\right) \overleftarrow{\times}_\theta \mathbb{Z}$$

where for $(a_0, \ldots, a_{n-1}, b), (c_0, \ldots, c_{n-1}, d) \in G_n$, $(a_0, \ldots, a_{n-1}, b) \leq (c_0, \ldots, c_{n-1}, d)$ iff $b < d$ or $b = d$ and $a_i \leq c_i$ for $0 \leq i \leq n-1$, and

$$(a_0, \ldots, a_{n-1}, b)(c_0, \ldots, c_{n-1}, d) = (a_0 + c_{0+b}, \ldots, a_{n-1} + c_{n-1+b}, b + d)$$

with all subscripts read modulo n. This ℓ-group may also be viewed as an ℓ-subgroup of $\mathbb{Z} \mathcal{W}r \mathbb{Z}$.

Letting $\mathcal{S}_n = \ell\text{-var}(G_n)$, i.e., the smallest variety of ℓ-groups containing G_n, it has been shown that $\mathcal{S}_n$ covers $\mathcal{A}$ in L for any prime number n [6]. This countably infinite family of covers can be translated, via the join operation in L, to produce a countably infinite family of covers over any variety $\mathcal{V}$ for which $\mathcal{V} \wedge \mathcal{S}_p = \mathcal{A}$ for p prime. For example, such families of covers lie over $\mathcal{R}$, $\mathcal{W}$, and each $\mathcal{S}_n$ and $\mathcal{L}_n$ (with the removal from the family of $\mathcal{S}_n \vee \mathcal{S}_p$ or $\mathcal{L}_n \vee \mathcal{S}_p$, where p is a prime factor of n). It is also readily seen, by considering the generating ℓ-groups, that for p and q distinct primes,

$$\mathcal{S}_p, \mathcal{S}_q \subset \mathcal{S}_p \vee \mathcal{S}_q \subset \mathcal{S}_{pq}$$

Thus, $\{\mathcal{S}_p | p \text{ prime}\}$ generates a sublattice of L isomorphic to the lattice of finite sets of primes. Yet every variety $\mathcal{S}_{p_1} \vee, \ldots, \vee \mathcal{S}_{p_n}$ in this sublattice is properly contained in a unique minimal Scrimger variety, e.g., $\mathcal{S}_{p_1 \cdots p_n}$. An attempt to examine the complete sublattice of L generated by $\{\mathcal{S}_p | p \text{ prime}\}$ led to the result below:

THEOREM 1.1. $\bigvee\{\mathcal{S}_{n_i} | i \in N\} = \mathcal{A}^2$, for any infinite subset $\{n_i\}$ of $N \backslash \{1\}$.

The proof of this theorem requires the following lemmas:

LEMMA 1.2. An ℓ-group G is ℓ-isomorphic to G_n iff G is generated by positive elements a, b where

(i) $a < b$, and

(ii) $b^{-k}ab^{k} \wedge a = 1$, for $1 \leq k < n$ and $b^{-n}ab^{n} = a$.

LEMMA 1.3. An ℓ-group G is ℓ-isomorphic to $\mathbb{Z}wr\mathbb{Z}$ iff G is generated by positive elements a,b where

(i) $a < b$, and

(ii) $b^{-k}ab^{k} \wedge a = 1$, for all $k \in N$.

Proof. Construct $G = \Pi\{G_{n_i} \mid i \in N\}$. Let $a = (a_{n_1}, a_{n_2}, \ldots)$ and $b = (b_{n_1}, b_{n_2}, \ldots)$, where a_{n_i}, b_{n_i} are the generators of G_{n_i} described in Lemma 1.2. Since for any fixed $k \in N$, only a finite number of the n_i will divide k,

$$b^{-k}ab^{k} \wedge a = (b_{n_1}^{-k}a_{n_1}b_{n_1}^{k} \wedge a_{n_1},\ b_{n_2}^{-k}a_{n_2}b_{n_2}^{k} \wedge a_{n_2}, \ldots)$$

has only a finite number of non-identity components. Thus, letting M be the ℓ-ideal of G generated by $\{b^{-k}ab^{k} \wedge a \mid k \in N\}$, a, b $\notin$ M and the cosets aM, bM generate an ℓ-subgroup of G/M ℓ-isomorphic to $\mathbb{Z}wr\mathbb{Z}$ by Lemma 1.3.

Another question of interest is whether or not $\mathcal{S}_n$ and $\mathcal{L}_n$ are distinct varieties. To answer this question for the case where n is not prime, let us inductively define

$$G_{n_1,\ldots,n_k} = \Big(\prod_{0}^{n_k-1} G_{n_1,\ldots,n_{k-1}}\Big) \overleftarrow{\times}_{\theta} \mathbb{Z}$$

for any finite collection of (not necessarily distinct) integers $n_1,\ldots,n_k$ from $N\backslash\{1\}$. Such an ℓ-group is again subdirectly irreducible, can be characterized in a manner similar to that in Lemma 1.2, and belongs to $\mathcal{A}^{k+1} \setminus \mathcal{A}^{k}$. Also, $G_{n_1,\ldots,n_k} \in \mathcal{L}_n$ iff $(n_1 \ldots n_k) \mid n$.

THEOREM 1.4. For n not prime, $\mathcal{S}_n \neq \mathcal{L}_n$.

Proof. Factor n as $n_1 \ldots n_k$ for some $k \geq 2$. Then $G_{n_1,\ldots,n_k} \in \mathcal{L}_n \setminus \mathcal{A}^k \subseteq \mathcal{L}_n \setminus \mathcal{A}^2$; while $\mathcal{S}_n \subseteq \mathcal{A}^2$. Thus, $G_{n_1,\ldots,n_k} \in \mathcal{L}_n \setminus \mathcal{S}_n$.

PROPOSITION 1.5. Let $\mathcal{S}_{n_1,\ldots,n_k} = \ell\text{-var}\,(G_{n_1,\ldots,n_k})$.

(i) $\mathcal{S}_{m_1,\ldots,m_k} \subseteq \mathcal{S}_{n_1,\ldots,n_k}$ iff $m_i \mid n_i$ for $1 \leq i \leq k$ (e.g. $\mathcal{S}_{2,3} \neq \mathcal{S}_{3,2}$).

(ii) $\mathcal{S}_{n_i} \subseteq \mathcal{S}_{n_1,\ldots,n_k}$ for $1 \leq i \leq k$.

(iii) $\mathcal{S}_{n_i,\ldots,n_{i+j}} \subseteq \mathcal{S}_{n_1,\ldots,n_k}$ for $1 \leq i \leq k$ and $0 \leq j \leq k-i$.

(iv) $\mathcal{S}_{(n_1\cdots n_k)} \subseteq \mathcal{S}_{n_1,\ldots,n_k}$.

Using this proposition, some results concerning varieties generated by wreath product ℓ-groups [2], and techniques similar to those employed in the proof of Theorem 1.1, it is possible to prove the following:

THEOREM 1.6.

(i) $\bigvee\{\mathcal{S}_{m,n(\lambda)} | \lambda \in N\} = \ell\text{-var}\ (G_m \mathit{wr} \mathbb{Z}) = \mathcal{S}_m\mathcal{A}$, whenever $\{n(\lambda)\}$ is an infinite subset of $N\setminus\{1\}$.

(ii) $\bigvee\{\mathcal{S}_{m(\lambda),n} | \lambda \in N\} = \ell\text{-var}\ (\mathbb{Z}\mathit{wr} G_n)$, whenever $\{m(\lambda)\}$ is an infinite subset of $N\setminus\{1\}$.

(iii) $\bigvee\{\mathcal{S}_{m(\lambda),n(\lambda)} | \lambda \in N\} = \ell\text{-var}\ (\mathbb{Z}\mathit{wr}\mathbb{Z}\mathit{wr}\mathbb{Z}) = \mathcal{A}^3$, whenever $\lambda \neq \gamma$ implies $m(\lambda) \neq m(\gamma)$ and $n(\lambda) \neq n(\gamma)$.

(iv) $\bigvee\{\mathcal{S}_{n_1,\ldots,n_i,\ n_{i+1}(\lambda),\ldots,n_k(\lambda)} | \lambda \in N\} = \ell\text{-var}\ (G_{n_1,\ldots,n_i} \mathit{wr}^{k-i}\mathbb{Z}) = \mathcal{S}_{n_1,\ldots,n_i}\mathcal{A}^{k-i}$, whenever $\lambda \neq \gamma$ implies $n_j(\lambda) \neq n_j(\gamma)$ for $i+1 \leq j \leq k$.

(v) $\bigvee\{\mathcal{S}_{n_1(\lambda),\ldots,n_k(\lambda)} | \lambda \in N\} = \ell\text{-var}\ (\mathit{wr}^{k+1}\mathbb{Z}) = \mathcal{A}^{k+1}$, whenever $\lambda \neq \gamma$ implies $n_j(\lambda) \neq n_j(\gamma)$ for $1 \leq j \leq k$.

THEOREM 1.7. $\bigvee\{\mathcal{L}_{n_i} | i \in N\} = \mathcal{N}$, whenever $\{\mathcal{L}_{n_i}\}$ is a properly nested family, i.e., $\mathcal{L}_{n_i} \subset \mathcal{L}_{n_{i+1}}$.

Proof. Recall that $\mathcal{L}_{n_i} \subset \mathcal{L}_{n_{i+1}}$ iff $n_i | n_{i+1}$ and $n_i \neq n_{i+1}$. Thus, $\{\mathcal{S}_{n_i}\}$ is an infinite collection of Scrimger varieties with each $\mathcal{S}_{n_i} \subseteq \mathcal{L}_{n_i}$, and so by Theorem 1.1, $\bigvee\mathcal{S}_{n_i} = \mathcal{A}^2 \subseteq \bigvee\mathcal{L}_{n_i}$. Similarly, $\{\mathcal{S}_{n_1,\ n_2/n_1}, \mathcal{S}_{n_2,\ n_4/n_2},\ldots\}$ is a collection of varieties for which $\mathcal{S}_{n_i,\ n_{2i}/n_i} \subseteq \mathcal{L}_{n_{2i}}$. So by Theorem 1.6 (iii), $\bigvee\mathcal{S}_{n_i,\ n_{2i}/n_i} = \mathcal{A}^3 \subseteq \bigvee\mathcal{L}_{n_{2i}} \subseteq \bigvee\mathcal{L}_{n_i}$. By continuing this process inductively, we can conclude that $\mathcal{A}^k \subseteq \bigvee\mathcal{L}_{n_i}$ of any $k \in N$; and thus, $\mathcal{N} = \bigvee\{\mathcal{A}^k | k \in N\} \subseteq \bigvee\{\mathcal{L}_{n_i} | i \in N\}$.

From this result it can be seen that an attempt to define additional new varieties $\mathcal{S}_{\{n_i\}}$ by use of direct limits fails, i.e., $\mathcal{S}_{\{n_i\}} = \mathcal{N}$ for any sequence $\{n_i\}$ from $N\setminus\{1\}$.

The work done thus far with $\mathcal{S}_{n_1,\ldots,n_k}$ varieties has relied primarily on our knowledge of their generating ℓ-groups. Little has been said of the equations that define these varieties. In fact even the question of what equations define $\mathcal{S}_n$ is still open. A few facts concerning this subject have been brought to light, however.

It has been shown [1] that $\mathcal{S}_p \neq \mathcal{L}_p$ for p prime by constructing an ℓ-group in $\mathcal{L}_p$ which satisfies

$$[[a,b,b]^{1+b+\ldots+b^{p-1}},c] = 1$$

and by showing that G_p does not satisfy this equation. (NOTE: $[a,b,c] = [[a,b], c]$ and $a^b = b^{-1}ab$)

Equations that are satisfied in G_n for $n \geq 2$ include:

(i) $[a^n, b^n] = 1$

(ii) $[a^{1+b+\ldots+b^{n-1}}, c^{1+d+\ldots+d^{n-1}}] = 1$

(iii) $[[a,b], [c,d]] = 1$

(iv) $[a,b,c^n] = 1$

(v) $[a,b,c]^{1+c+\ldots+c^{n-1}} = 1$

(vi) $[a, kb]^{1+b+\ldots+b^{n-1}} = 1$ for any $k \geq 2$.

(vii) $[a^{1+b_1+\ldots+b_{n-1}}, c^{1+d_1+\ldots+d_{n-1}}] = 1.$

However, there is a good deal of interdependence among the equations in this list. For example, equations (i) and (ii) are equivalent, as are equations (iv) and (v), equation (v) implies (vi), and equation (vii) implies all of the equations in the list. Thus, the variety defined by equation (vii) is a subvariety of $\mathcal{L}_n$ (defined by (i)); but is it a proper subvariety? If $n = m_1m_2$, then equation (iv) is not satisfied in $\mathcal{S}_{m_1,m_2}$. So equation (iv) cannot define $\mathcal{L}_n$, but it does define a variety containing $\mathcal{S}_n$. Does it define $\mathcal{S}_n$? On the whole, the connection between the $\mathcal{S}_{n_1,\ldots,n_k}$ varieties and equations is yet unknown.

REFERENCES

1. C. D. Fox, On the Scrimger varieties of lattice ordered groups. In press.

2. A. M. W. Glass, W. C. Holland, and S. H. McCleary, The structure of ℓ-group varieties. In press, in *Algebra Universalis*.

3. W. C. Holland, The largest proper variety of lattice ordered groups. *Proc. Amer. Math. Soc.* 57: 25-28 (1976).

4. J. Martinez, Free products in varieties of lattice-ordered groups. *Czech. Math. J.* 22(97): 535-553 (1972).

5. J. Martinez, Varieties of lattice-ordered groups. *Math. Zeitschr.* 137: 265-284 (1974).

6. E. B. Scrimger, A large class of small varieties of lattice-ordered groups. *Proc. Amer. Math. Soc.* 51: 301-306 (1975).

7. J. E. H. Smith, The lattice of ℓ-group varieties. *Trans. Amer. Math. Soc.* 257: 347-357 (1980).

8. E. C. Weinberg, Free lattice-ordered abelian groups II, *Math. Ann.* 159: 217-222 (1965).

A SOLUTION OF THE WORD PROBLEM IN FREE NORMAL-VALUED LATTICE-ORDERED GROUPS

Stephen H. McCleary

University of Georgia
Athens, Georgia

1. INTRODUCTION

In any lattice-ordered group (ℓ-group) generated by a set X, every element can be written (not uniquely) in the form $w(\underset{\sim}{x}) = \bigvee_i \bigwedge_j w_{ij}(\underset{\sim}{x})$, where each $w_{ij}(\underset{\sim}{x})$ is a group word in the elements of X. An algorithm will be given for deciding whether $w(\underset{\sim}{x})$ is the identity e in the free normal valued ℓ-group on X, or equivalently, whether the statement "$\forall \underset{\sim}{x},\ w(\underset{\sim}{x}) = e$" holds in all normal valued ℓ-groups. The algorithm is quite different from the one given recently by Holland and McCleary [6] for the free ℓ-group, and indeed the solvability of the word problem was established first for the normal valued case. The present algorithm makes crucial use of the fact (due to Glass, Holland, and McCleary [3]) that the variety of normal valued ℓ-groups is generated by the finite wreath powers $\mathbb{Z}\mathrm{Wr}\mathbb{Z}\mathrm{Wr}\ldots\mathrm{Wr}\mathbb{Z}$ of the integers $\mathbb{Z}$.

Many natural statements about ℓ-groups involve variables which are constrained to be positive. We shall give algorithms to decide whether

$$\forall \underset{\sim}{x} \geq \underset{\sim}{e},\ \forall \underset{\sim}{y},\ w(\underset{\sim}{x},\underset{\sim}{y}) \geq e,$$

$$\forall \underset{\sim}{x} \geq \underset{\sim}{e},\ \forall \underset{\sim}{y},\ w(\underset{\sim}{x},\underset{\sim}{y}) \leq e,$$

and $\forall \underset{\sim}{x} \geq \underset{\sim}{e},\ \forall \underset{\sim}{y},\ w(\underset{\sim}{x},\underset{\sim}{y}) = e$

hold in all normal valued ℓ-groups.

---

*This is an expanded version of material developed while the author was on leave at Bowling Green State University in Bowling Green, Ohio.

Normal valued ℓ-groups can be defined in many ways, e.g., as those ℓ-groups satisfying any one of the following equivalent laws [2]:

$$\forall x, z \geq e, xz \leq z^2x^2$$
$$\forall x, z \geq e, xz \leq z^m x^n \text{ (for any fixed m, n} \geq 2)$$
$$\forall x, z \geq e, (x^{-1}z^{-1}xz)^n \leq x \vee z \text{ (for any fixed n} \geq 2)$$
$$\forall x, z, (x^{-1}z^{-1}xz)^n \leq |x| \vee |z| \text{ (for any fixed n} \geq 2)$$
$$\forall x, z \geq e, (x^2z^{-2} \wedge x^{-2}z^2)^n \leq x \wedge z \text{ (for any fixed n} \geq 1)$$
$$\forall x, z \geq e, (x^2z^{-2} \wedge x^{-2}z^2)^n \leq x \text{ (for any fixed n} \geq 1)$$

Here $|x|$ means $x \vee x^{-1}$. Each of these is easily translated into a bona fide law. For example, the first translates to

$$\forall x, z, (x \vee e)(z \vee e)(x \vee e)^{-2}(z \vee e)^{-2} \vee e = e$$

In attempting to enlarge this list of defining laws for the variety $\mathcal{N}$ of normal valued ℓ-groups, one would like to be able to decide whether a proposed law does indeed define $\mathcal{N}$. Together with the decision procedure of Holland and McCleary [6] for ℓ-groups, the present paper provides such a capability. For since $\mathcal{N}$ is the unique largest proper variety of ℓ-groups [4], the statement "$\forall \underset{\sim}{x} \geq \underset{\sim}{e}, \forall \underset{\sim}{y}, w(\underset{\sim}{x},\underset{\sim}{y}) = e$" defines $\mathcal{N}$ if and only if it holds in (all ℓ-groups in) $\mathcal{N}$ but does not hold in all ℓ-groups.

It will be essential for the reader to have a good understanding of the wreath product $\mathbb{Z}\text{Wr}\mathbb{Z}$ as an ℓ-permutation group. A bit more generally, we define the wreath product $(G,S)\text{Wr}(\mathbb{Z},\mathbb{Z})$ of an arbitrary ℓ-permutation group with the regular representation of $\mathbb{Z}$. Let R be the totally ordered set $\overleftarrow{S \times \mathbb{Z}}$ ordered lexicographically from the right. Let $W = \{(\{g_m | m \in \mathbb{Z}\}, z)\}$, where each $g_m \in G$ and $z \in \mathbb{Z}$. We write (g_m,z) in place of $(\{g_m|m \in \mathbb{Z}\}, z)$. The g_m's are called the *local components* of (g_m,z), and z is called the *global component*. Now let W act on R in the natural way, i.e., $(s,t)(g_m,z) = (sg_t, t + z)$. This makes (W,R) an ℓ-permutation group, known as the *wreath product* $(G,S)\text{Wr}(\mathbb{Z},\mathbb{Z})$. The resulting group operation on W is given by

$$(g_m, z)(f_m, y) = (g_m f_{m+z}, z + y),$$
$$\text{and } (g_m,z)^{-1} = (g_{m-z}^{-1}, -z).$$

The order is given by: $(g_m, z) \geq e$ if and only if either $z > 0$, or $z = 0$ and each $g_m \geq e$. When $h \in W$, we define the notations $\hat{h}$ and h_m so that $h = (h_m,\hat{h})$.

The ℓ-group $(G,S)\mathrm{Wr}(\mathbb{Z},\mathbb{Z})$ is independent of the representation (G,S) of G, and for $(\mathbb{Z},\mathbb{Z})$ we write simply $\mathbb{Z}$, so that the ℓ-group will be denoted by $G\mathrm{Wr}\mathbb{Z}$. The nth wreath power $\mathrm{Wr}^n\mathbb{Z}$ means $(\mathrm{Wr}^{n-1}\mathbb{Z})\mathrm{Wr}\mathbb{Z}$. We shall denote the point $(0,\ldots,0)$ by $\underset{\sim}{0}$.

We shall make crucial use of the fact that $\mathcal{N}$ is generated by the set of finite wreath powers $\mathrm{Wr}^n\mathbb{Z}$ of $\mathbb{Z}$ (Glass, Holland, and McCleary [3]). Equivalently, " $\forall \underset{\sim}{x} \geq \underset{\sim}{e}$, $\forall \underset{\sim}{y}$, $w(\underset{\sim}{x},\underset{\sim}{y}) = e$" *holds in* $\mathcal{N}$ *if and only if it holds in all finite wreath powers of* $\mathbb{Z}$; and similarly if "=" is replaced by "$\geq$" or "$\leq$".

George McNulty has pointed out that since the restricted finite wreath powers of $\mathbb{Z}$ (restricted so that each (g_m, z) has only finitely many non-zero g_m's) also generate $\mathcal{N}$ and are uniformly recursively enumerable, standard logical techniques can be applied to show from the above statement that the word problem is solvable in free normal valued ℓ-groups. (Briefly, one machine searches through the wreath powers for a counterexample while another generates all possible proofs.) Here, however, we seek algorithms with some hope of implementation in the real world. In fact, the ones presented here run in non-deterministic polynomial time.

Our algorithms will check whether a statement holds in the successive wreath powers $\mathrm{Wr}^n\mathbb{Z}$. The nth "stage" of the algorithm will deal with the nth wreath power, and unless a negative answer is obtained there, will manufacture a new word to be processed in the next stage. In each stage, we shall need information about integer solutions of systems of simultaneous linear inequalities. In §5, we shall discuss an algorithm (due to Motzkin [9]) for finding all solutions of such a system. §3 and §5 are enough to establish the solvability of the word problem if efficiency is not an object.

In fact, for many words it is unnecessary in practice to use the Motzkin Algorithm, either because the systems of inequalities are simple enough to solve otherwise, or because they yield to the non-algorithmic approach of §7.

These investigations yield a curious by-product: in the free normal valued ℓ-group on a set y, any element which can be put in standard form so that no group word w_{ij} is e has trivial polar. This holds also for (absolutely) free ℓ-groups.

Let us look at an example to get a preliminary idea of what the algorithm will do to show that a statement about $w(x,z)$ holds in $\mathcal{N}$. We consider

$$\forall x,\ z \geq e,\ xxz^{-1} \vee zx^{-1}x^{-1}zx^{-1}x^{-1} \vee xz^{-1} \geq e.$$

(Admittedly it is clear by inspection that this statement holds in all ℓ-groups, but it will still serve as a good illustration.) We look at the abelianized group words: $2x - z$, $-4x + 2z$, $x - z$. It is easily seen that in the integers $\mathbb{Z}$, the three abelianized words cannot be simultaneously negated, so that the statement holds in $\mathbb{Z}$.

In order to make the statement fail in $\mathbb{Z}\mathrm{Wr}\mathbb{Z}$, we would need to choose the global components $\hat{x}$ and $\hat{z}$ of x and z so that they were non-negative, and so that they made the abelianized words non-positive. This can be done so as to make $x - z < 0$, so we suppress the third word. Making the remaining abelianized words non-positive requires that $2x - z = 0$. This in turn puts conditions on the local components used for (0,0) in applying the various occurrences of x and z. The local components must be the same for the first, fourth, and sixth occurrences of x, for the second, third, and fifth occurrences of x, and for all three occurrences of z; and these three local components can be chosen arbitrarily. Accordingly, we "split" the occurrences of x and consider a statement about a new word $\bar{w}$:

$$\forall x_1, x_2, z, \; x_1x_2z^{-1} \vee zx_2^{-1}x_1^{-1}zx_2^{-1}x_1^{-1} \geq e.$$

Because the three abelianized words could be made non-positive with $x > 0$ and $z > 0$, the variables are now unconstrained. At this point, we have finished the first stage of the algorithm.

We treat $\bar{w}$ as we did w, and look at the abelianized words $x_1 + x_2 - z$ and $-2x_1 - 2x_2 + z$. They cannot be simultaneously negated in $\mathbb{Z}$, so the statement about $\bar{w}$ holds in $\mathbb{Z}$ and thus the original statement about w holds in $\mathbb{Z}\mathrm{Wr}\mathbb{Z}$. In order to make the statement about $\bar{w}$ fail in $\mathbb{Z}\mathrm{Wr}\mathbb{Z}$ (and thus make the statement about w fail in $\mathbb{Z}\mathrm{Wr}\mathbb{Z}\mathrm{Wr}\mathbb{Z}$), we would need to choose the global components of x_1, x_2 and z so as to make the new abelianized words non-positive, thus making $x_1 + x_2 - z = 0$, which would put conditions on the local components that would cause there to be no further splitting of variables. The fact that the word thus created from $\bar{w}$ is $\bar{w}$ itself causes the algorithm to terminate (in the second stage) and guarantees that the original statement does indeed hold in $\mathcal{N}$.

When one of our algorithms decides that a statement fails in $\mathcal{N}$, the data generated during its execution can be used to write down a specific substitution in some $\mathrm{Wr}^n\mathbb{Z}$ demonstrating this failure, thereafter permitting one to prove failure by merely exhibiting this substitution. Unfortunately, no such convenient device is available when the decision is positive.

2. NOTATION

Let $X = \{x_1, x_2, \ldots\}$ be a countably infinite set. Each ℓ-group word can be written as $w(\underset{\sim}{x}) = \bigvee_{i\in I} w_i(\underset{\sim}{x}) = \bigvee_{i\in I} \bigwedge_{j\in J_i} \prod_{k\in K_{ij}} x_{ijk}^{n_{ijk}}$. Here the index sets are finite sets of the form $\{1, 2, \ldots, m\}$, each $x_{ijk} \in X \cup \{e\}$, each $n_{ijk} = \pm 1$ and each group word w_{ij} is in reduced form as an element of the free group on X. The w_i's will be called the *terms* of w. When we speak of the *occurrences* of a given x, we include those with exponent -1.

Usually some of the variables $\underset{\sim}{x}$ will be constrained to be non-negative, and others $\underset{\sim}{y}$ will be unconstrained, as in

$$\forall \underset{\sim}{x} \geq \underset{\sim}{e}, \forall \underset{\sim}{y}, \; w(\underset{\sim}{x},\underset{\sim}{y}) = e.$$

To say that this statement holds in all normal valued ℓ-groups is equivalent to saying that $w(\underset{\sim}{x} \vee \underset{\sim}{e}, \underset{\sim}{y}) = e$ in the free normal valued ℓ-group on $\underset{\sim}{x} \cup \underset{\sim}{y}$, where $\underset{\sim}{x} \vee \underset{\sim}{e}$ denotes the sequence of elements $x \vee e$ $(x \in \underset{\sim}{x})$. Of course, a variable z constrained to be non-positive can be dealt with by rewriting the word in terms of z^{-1} and constraining z to be non-negative.

We shall frequently be interested in whether a group word $w_{ij}(\underset{\sim}{x},\underset{\sim}{y})$ is e, and there is a potential ambiguity here, which fortunately is resolved by the following lemma guaranteeing that if "$\forall \underset{\sim}{x} \geq \underset{\sim}{e}, \forall \underset{\sim}{y}, w_{ij} \geq e$" holds in $\mathcal{N}$, w_{ij} must be formally e. Also, for many ℓ-group words w, the lemma lets one tell by inspection that "$\forall \underset{\sim}{x} \geq \underset{\sim}{e}, \forall \underset{\sim}{y}, w = e$" fails in $\mathcal{N}$.

LEMMA 1. For an ℓ-group word $w = \bigvee_i \bigwedge_j w_{ij}$, if "$\forall \underset{\sim}{x} \geq \underset{\sim}{e}, \forall \underset{\sim}{y}, w(\underset{\sim}{x},\underset{\sim}{y}) = e$" holds in $\mathcal{N}$, then at least one w_{ij} is formally e. In particular, if "$\forall \underset{\sim}{x} \geq \underset{\sim}{e}, \forall \underset{\sim}{y}, w_{ij}(\underset{\sim}{x},\underset{\sim}{y}) = e$" holds in $\mathcal{N}$ for a group word w_{ij}, then w_{ij} is formally e.

Proof. The free group on $\underset{\sim}{x} \cup \underset{\sim}{y}$ can be totally ordered in such a way that each generator has a predesignated sign [1], and is then in $\mathcal{N}$. In this ordering, $w = \max_i \min_j w_{ij} \neq e$.

Let G be an ℓ-group. A substitution for an ℓ-group word $w(\underset{\sim}{x},\underset{\sim}{y})$ in G is a function $g: \underset{\sim}{x} \cup \underset{\sim}{y} \cup \{e\} \to G$ such that $g(e) = e$ and for each $x \in \underset{\sim}{x}$, $g(x) \geq e$. By $w(g)$ we mean $\bigvee_i \bigwedge_j \prod_k g(x_{ijk})^{n_{ijk}}$.

We mention a bit of background information which will permit more graceful statements of some secondary results. Let $\mathcal{A}$ denote the variety of abelian ℓ-groups. For the product varieties $\mathcal{A}^n$, we have $\mathcal{A} \subset \mathcal{A}^2 \subset \ldots \subset \mathcal{A}^n \subset \ldots \subset \mathcal{N}$, and $\mathcal{A}^n$ is the variety generated by $\mathrm{Wr}^n\mathbb{Z}$ [3, Theorem 4.6]. (Product varieties are defined in [3], but the reader unfamiliar with them will lose little if he pretends that $\mathcal{A}^n$ is merely a peculiar notation for the variety generated by $\mathrm{Wr}^n\mathbb{Z}$.) At any rate, en route to a decision on whether a statement holds in $\mathcal{N}$, the algorithms decide whether it holds in the successive wreath powers $\mathrm{Wr}^n\mathbb{Z}$, and thus give decision procedures for the varieties $\mathcal{A}^n$.

3. THE STATEMENT "$\forall\, \underset{\sim}{x} \geq \underset{\sim}{e},\ \forall\, \underset{\sim}{y},\ w(\underset{\sim}{x},\underset{\sim}{y}) \geq e$"

The distributive laws give

PROPOSITION 2. Let $\mathcal{V}$ be any variety, and let $w = \bigvee_{i\in I} \bigwedge_{j\in J_i} w_{ij}$. "$\forall\, \underset{\sim}{x} \geq \underset{\sim}{e},\ \forall\, \underset{\sim}{y},\ w(\underset{\sim}{x},\underset{\sim}{y}) \geq e$" holds in $\mathcal{V}$ if and only if "$\forall\, \underset{\sim}{x} \geq \underset{\sim}{e},\ \forall\, \underset{\sim}{y},\ \bigvee_{i\in I} w_{ij_i}(\underset{\sim}{x},\underset{\sim}{y}) \geq e$" holds in $\mathcal{V}$ for every choice of j_i's with no $w_{ij_i} = e$.

In practice, it is convenient to modify w by erasing all group words w_{ij} which are e, and then to test the modified word. (If this erasure causes an entire term $w_i = \bigwedge_j w_{ij}$ to disappear, then of course "$\forall\, \underset{\sim}{x} \geq \underset{\sim}{e},\ \forall\, \underset{\sim}{y},\ w(\underset{\sim}{x},\underset{\sim}{y}) \geq e$" holds automatically.)

Let $\mathcal{S}(w)$ be the set of words $\dot{w}(\dot{x},\dot{y}) = \bigvee_{i\in I} w_{ij_i}$ (with no $w_{ij_i} = e$) associated with w, which unfortunately can be fairly large. We proceed through $\mathcal{S}(w)$, deciding the above statement for one $\dot{w}$ at a time until either some $\dot{w}$ gives a negative answer or all have given affirmative answers. Of course, if one suspects a negative answer, one may try to be (non-algorithmically) clever about the choice of $\dot{w}$.

Now we specialize to $\mathcal{N}$. For a given $\dot{w}$, the following Algorithm 9 is not always needed--the answer may be obvious. Also, when the algorithm for "$\forall\, \underset{\sim}{x} \geq \underset{\sim}{e},\ \forall\, \underset{\sim}{y},\ w(\underset{\sim}{x},\underset{\sim}{y}) \geq e$" is being used to help decide whether "$\forall \underset{\sim}{x} \geq \underset{\sim}{e},\ \forall\, \underset{\sim}{y},\ w(\underset{\sim}{x}\ \underset{\sim}{y}) = e$", drastic reductions in the number of $\dot{w}$'s can often be effected; cf. §4.

Now we turn our attention to words of the form $w(\underset{\sim}{x},\underset{\sim}{y}) = \bigvee_{i\in I} w_i = \bigvee_{i\in I} \prod_k x_{ik}^{n_{ik}}$ with no $w_i = e$, and construct the algorithm for deciding whether "$\forall\, \underset{\sim}{x} \geq \underset{\sim}{e},\ \forall\, \underset{\sim}{y},\ w(\underset{\sim}{x},\underset{\sim}{y}) \geq e$" holds in $\mathcal{N}$.

For each $w_i(\underset{\sim}{x},\underset{\sim}{y})$, we let $C_i = C(w_i)$ denote the abelianization of w_i written additively; i.e., the formal sum $\Sigma\, a_q z_q$, where z_q ranges over the letters of X involved in $w_i(\underset{\sim}{x},\underset{\sim}{y})$ and a_q is the sum of the exponents for the occurrences of z_q in w_i. When the a_q's are not all 0, the equation $C_i = 0$ represents a hyperplane in d(w)-dimensional rational space, where $d(x) = |\underset{\sim}{x} \cup \underset{\sim}{y}|$.

We shall be interested in systems involving various linear equations $C = 0$ and inequalities $C \leq 0$ and $C < 0$ and $C \neq 0$ in which all coefficients are integers. Our interest will actually be in integer solutions, but when solving the systems we shall deal with rational solutions. If the computations are done by computer, they must be done using rational arithmetic.

ALGORITHM 3. (for the abelian variety $\mathcal{A}$). Let $w = \bigvee_{i\in I} w_i$. "$\forall\, \underset{\sim}{x} > \underset{\sim}{e},\ \forall\, \underset{\sim}{y},\ w(\underset{\sim}{x},\underset{\sim}{y}) \geq e$" fails in $\mathcal{A}$ if and only if the system

$$(1)\qquad \begin{array}{ll} x \geq 0 & (\text{for each } x \in \underset{\sim}{x}) \\ C_i < 0 & (\text{for each } i \in I) \end{array}$$

has an integer solution.

An algorithm for deciding whether (1) has an integer (equivalently rational) solution will be given in §5, for use when no easier method presents itself. That the word problem is solvable in $\mathcal{A}$ is not new, being due to N. G. K. Kisamov [7].

If "$\forall\, \underset{\sim}{x} \geq \underset{\sim}{e},\ \forall\, \underset{\sim}{y},\ w(\underset{\sim}{x},\underset{\sim}{y}) \geq e$" fails in $\mathcal{A}$, then of course it fails in $\mathcal{N}$. Now suppose it holds in $\mathcal{A}$. We shall manufacture from $w(\underset{\sim}{x},\underset{\sim}{y})$ another ℓ-group word $\bar{w}(\bar{\underset{\sim}{x}},\bar{\underset{\sim}{y}})$ which will in turn be tested in $\mathbb{Z}$ to determine whether the original statement about w holds in $\mathbb{Z}\mathrm{Wr}\mathbb{Z}$.

Consider the system

$$(2)\qquad \begin{array}{ll} x \geq 0 & (x \in \underset{\sim}{x}), \\ C_i \leq 0 & (i \in I). \end{array}$$

Let $\underset{\sim}{x}_0 = \{x \in \underset{\sim}{x} \mid (2) \text{ implies } x = 0\}$, the implication referring to integral, or equivalently rational, solutions. Let $I_0 = \{i \in I \mid (2) \text{ implies } C_i = 0\}$, which of course includes all $i \in I$ for which the abelianization C_i has all coefficients equal to zero. Since "$\forall\, \underset{\sim}{x} \geq \underset{\sim}{e},\ \forall\, \underset{\sim}{y},\ w(\underset{\sim}{x},\underset{\sim}{y}) \geq e$" holds in $\mathbb{Z}$, I_0 is non-empty. Algorithms for finding $\underset{\sim}{x}_0$ and I_0 will be given in §5. For each $x \in \underset{\sim}{x} \setminus \underset{\sim}{x}_0$, there exists an integer solution of (2) for which $x > 0$; and for each $i \in I \setminus I_0$, there exists an integer solution of (2) for which

$C_i < 0$. Adding these solutions, we obtain a solution of

$$(3)\qquad \begin{aligned} x &= 0 \quad (x \in \underset{\sim}{x}_0),\\ x &> 0 \quad (x \in \underset{\sim}{x} \setminus \underset{\sim}{x}_0),\\ C_i &= 0 \quad (i \in I_0),\\ C_i &< 0 \quad (i \in I \setminus I_0). \end{aligned}$$

Let $\tilde{w}(\underset{\sim}{x},\underset{\sim}{y}) = \bigvee_{i \in I_0} w_i = \bigvee_{i \in I_0} \prod_k x_{ik}^{n_{ik}}$

We shall form $\bar{w}$ from $\tilde{w}$ by "splitting the variables" of $\tilde{w}$. The intent is this: If $x_{i_1k_1}$ and $x_{i_2k_2}$ are the same variable (same element of X), we want them to remain the same variable in $\bar{w}$ if and only if for every substitution $z \to g(z)$ for $\tilde{w}$ in $\mathbb{Z}\mathrm{Wr}\mathbb{Z}$ whose global components $\hat{g}(z)$ satisfy (3), the same local component $g(x_{i_1k_1})_m$ of $g(x_{i_1k_1})$ is applied for $\mathbb{Z}_0 = \{(m,0) \mid m \in \mathbb{Z}\}$ in the processing $x_{i_1k_1}^{\pm 1}$ as in the processing of $x_{i_2k_2}^{\pm 1}$.

Let C_{ik} denote the abelianization $C(x_{i1}^{n_{i1}} \ldots x_{ik}^{n_{ik}})$; or if $k = 0$, let $C_{ik} = 0$. Let D_{ik} be

$$\begin{aligned} &C_{i,k-1} && \text{if } n_{ik} = +1,\\ &C_{ik} && \text{if } n_{ik} = -1. \end{aligned}$$

Let $z \to g(z)$ be any substitution for w in $\mathbb{Z}\mathrm{Wr}\mathbb{Z}$. Let $p = C_{i,k-1}(\underset{\sim}{\hat{g}}(z))$, the result of substituting the global component $\hat{g}(z)$ for each z in $C_{i,k-1}$. Let $q = D_{ik}(\underset{\sim}{\hat{g}}(z)) =$

$$\begin{aligned} &p && \text{if } n_{ik} = +1,\\ &p - \hat{g}(x_{ik}) && \text{if } n_{ik} = -1. \end{aligned}$$

When $x_{ik}^{n_{ik}}$ is reached in the processing of w_i, $\mathbb{Z}_0$ has been moved to $\mathbb{Z}_p = \{(m,p) \mid m \in \mathbb{Z}\}$; and the integer then applied locally is

$$\begin{aligned} &g(x_{ik})_q && \text{if } n_{ik} = +1,\\ -&g(x_{ik})_q && \text{if } n_{ik} = -1. \end{aligned}$$

Thus $D_{ik}(\underset{\sim}{\hat{g}}(z)) = q$ selects the relevant local component of $g(x_{ik})$.

We want two occurrences $x_{i_1k_1}$ and $x_{i_2k_2}$ of the same variable in w

to remain the same variable in $\bar{w}$ if and only if $D_{i_1 k_1}(\hat{g}(z)) = D_{i_2 k_2}(\hat{g}(z))$
for every substitution $z \to g(z)$ for w in $\mathbb{Z}Wr\mathbb{Z}$ whose global components $\hat{g}(z)$
satisfy (3); or more simply, if and only if (3) implies $D_{i_1 k_1} = D_{i_2 k_2}$,
where the implication refers to integer (equivalently, rational) solutions.

Determining whether (3) implies $D_{i_1 k_1} = D_{i_2 k_2}$ is easy with the aid of

$$(4) \quad \begin{aligned} x &= 0 \quad (x \in x_0), \\ C_0 &= 0 \quad (i \in I_0). \end{aligned}$$

LEMMA 4. The following are equivalent:

(a) (3) implies $D_{i_1 k_1} = D_{i_2 k_2}$.

(b) (4) implies $D_{i_1 k_1} = D_{i_2 k_2}$.

(c) (2) implies $D_{i_1 k_1} = D_{i_2 k_2}$.

Proof. Clearly (b) implies (a), and by the definitions of x_0 and I_0,
(c) implies (b). We show (a) implies (c). Suppose $\underset{\sim}{a}$ is an integer solution
of (2) for which some $D_{i_1 k_1}(\underset{\sim}{a}) \neq D_{i_2 k_2}(\underset{\sim}{a})$. By the remarks prior to (3), we
may pick an integer solution $\underset{\sim}{b}$ of (3). Then for a sufficiently large n,
$\underset{\sim}{a} + n\underset{\sim}{b}$ is an integer solution of (3) for which $D_{i_1 k_1}(\underset{\sim}{a} + n\underset{\sim}{b}) \neq D_{i_2 k_2}(\underset{\sim}{a} + n\underset{\sim}{b})$.

ALGORITHM 5. (for deciding whether to split two occurrences
$x_{i_1 k_1}$ and $x_{i_2 k_2}$ of the same variable in $\tilde{w}$). Let W be the solution space
for (4) in $d(w)$-dimensional rational space. Find a basis F for W, arranging
if desired that each $\underset{\sim}{f} \in F$ be an integer vector. Then $x_{i_1 k_1}$ and $x_{i_2 k_2}$
remain the same variable in $\bar{w}$ if and only if $D_{i_1 k_1}(\underset{\sim}{f}) = D_{i_2 k_2}(\underset{\sim}{f})$ for all
$\underset{\sim}{f} \in F$.

An alternative to this algorithm will be given in §5.

For two occurrences x_{ik_1} and x_{ik_2} $(k_1 < k_2)$ of the same variable in
the same w_i, the splitting criterion is simpler: They remain the same
variable in $\bar{w}$ if and only if (4) implies

$$\begin{aligned} C(x_{ik_1}^{\pm 1} \cdots x_{i,k_2-1}^{\pm 1}) &= 0 \text{ when } n_{ik_1} = n_{ik_2}, \\ C(x_{i,k_1+1}^{\pm 1} \cdots x_{i,k_2-1}^{\pm 1}) &= 0 \text{ when } n_{ik_1} = -n_{ik_2}. \end{aligned}$$

Proof. If $n_{ik_1} = +1$ and $n_{ik_2} = -1$, $C(x_{i,k_1}^{+1} \ldots x_{i,k_2}^{-1})$ $= C(x_{i,k_1+1}^{\pm 1} \ldots x_{i,k_2-1}^{\pm 1})$.

Now we form $\bar{w}(\bar{\underset{\sim}{x}},\bar{\underset{\sim}{y}})$ from $\tilde{w}(\underset{\sim}{x},\underset{\sim}{y})$ by *splitting the variables* as follows (cf. the above statement of intent): For each variable z occurring in $\tilde{w}$, we replace the occurrences of z by several variables from X in such a way that $x_{i_1k_1}$ and $x_{i_2k_2}$ are replaced by the same variable if and only if (4) implies $D_{i_1k_1} = D_{i_2k_2}$, and arrange that the occurrences of distinct z's are replaced by disjoint sets of variables. These "descendants" of z are constrained in $\bar{w}$ if $z \in \underset{\sim}{x}_0$, and are unconstrained if $z \in \underset{\sim}{x} \setminus \underset{\sim}{x}_0$ or $z \in \underset{\sim}{y}$. We write $\bar{w}(\bar{\underset{\sim}{x}},\bar{\underset{\sim}{y}})$, with $\bar{\underset{\sim}{x}}$ consisting of those variables still constrained.

We have now completed the first stage of the algorithm.

LEMMA 6. Let $w = \bigvee_{i\in I} w_i$, and suppose the statement "$\forall\, \underset{\sim}{x} \geq e$, $\forall\, \underset{\sim}{y}$, $w(\underset{\sim}{x},\underset{\sim}{y}) \geq e$" holds in $\mathbb{Z}$. Then this statement holds in $\mathbb{Z}\mathrm{Wr}\mathbb{Z}$ if and only if "$\forall\, \bar{\underset{\sim}{x}} \geq e$, $\forall\, \bar{\underset{\sim}{y}}$, $\bar{w}(\bar{\underset{\sim}{x}},\bar{\underset{\sim}{y}}) \geq e$" holds in $\mathbb{Z}$.

Proof. Since the original statement holds in $\mathbb{Z}$, the word $\bar{w}$ is defined.

First, let $z \to g(z)$ be a substitution for $w(\underset{\sim}{x},\underset{\sim}{y})$ in $\mathbb{Z}\mathrm{Wr}\mathbb{Z}$ for which $w(\underset{\sim}{g}) \not\geq e$. We shall produce a substitution $\bar{z} \to h(\bar{z})$ for $\bar{w}(\bar{\underset{\sim}{x}},\bar{\underset{\sim}{y}})$ in $\mathbb{Z}$ for which $\bar{w}(\underset{\sim}{h}) \not\geq e$.

Conjugating by an appropriate element of $\mathbb{Z}\mathrm{Wr}\mathbb{Z}$, we may assume that $(0,0)w(\underset{\sim}{g}) < (0,0)$. Since $w(\hat{\underset{\sim}{g}}) \geq e$ in $\mathbb{Z}$, $(0,0)w(\underset{\sim}{g}) = (\alpha,0)$ for some $\alpha < 0$. Let $\bar{I} = \{i \in I \mid w_i(\hat{\underset{\sim}{g}}) = 0\}$. Since $w_i(\hat{\underset{\sim}{g}}) \leq 0$ for each $i \in I$, we have $\bar{I} \supseteq I_0 = \{i \in I \mid (2) \text{ implies } C_i = 0\}$, the indexing set for $\tilde{w}$ and $\bar{w}$. Let $\bar{x} = \{x \in \underset{\sim}{x} \mid \hat{g}(x) = 0\} \supseteq \underset{\sim}{x}_0$.

If $x_{i_1k_1}$ and $x_{i_2k_2}$ remain the same variable when the variables are split to form $\bar{w}$ from $\tilde{w}$, then for every substitution for w in $\mathbb{Z}\mathrm{Wr}\mathbb{Z}$ for which the global components satsify (2), and thus in particular for the present substitution $z \to g(z)$, the same local component $g(x_{ik})_m$ of $g(x_{ik})$ is applied for $\mathbb{Z}_0$ in the processing of $x_{i_1k_1}^{\pm 1}$ as in the processing of $x_{i_2k_2}^{\pm 1}$. Hence $x_{ik} \to g(x_{ik})_m$ is a well defined substitution for $\bar{w}$ in $\mathbb{Z}$. (If x_{ik} is constrained in $\bar{w}$, then $x \in \underset{\sim}{x}_0 \subseteq \bar{\underset{\sim}{x}}$, so $\hat{g}(x_{ik}) = 0$, and since x_{ik} must have been constrained in w, we then have $g(x_{ik})_m \geq 0$.) We denote this substitution by $\bar{z} \to h(\bar{z})$. For $i \in I_0$, we have $(0,0)w_i(\underset{\sim}{g}) = (0w_i(\underset{\sim}{h}), 0)$.

Now we have

$(0,0) > (\alpha,0) = (0,0)w(\underset{\sim}{g}) = (0,0)(\bigvee_{i\in I} w_i(\underset{\sim}{g})) \geq (0,0)(\bigvee_{i\in I_0} w_i(\underset{\sim}{g}))$

$= \max_{i\in I_0}(0,0)w_i(\underset{\sim}{g}) = \max_{i\in I_0}(0\bar{w}_i(\underset{\sim}{h}),\ 0) = (\max_{i\in I_0} 0\bar{w}_i(\underset{\sim}{h}),\ 0)$

$= (0(\bigvee_{i\in I_0} \bar{w}_i(\underset{\sim}{h})),\ 0) = (0\bar{w}(\underset{\sim}{h}),\ 0)$. Hence $0\bar{w}(\underset{\sim}{h}) < 0$, so that h is a substitution showing that "$\forall\, \underset{\sim}{\bar{x}} \geq \underset{\sim}{e},\ \forall\, \underset{\sim}{\bar{y}},\ \bar{w}(\underset{\sim}{\bar{x}},\underset{\sim}{\bar{y}}) \geq e$" fails in $\mathbb{Z}$.

Conversely, let $\bar{z} \to h(\bar{z})$ be a substitution for $\bar{w}(\underset{\sim}{\bar{x}},\underset{\sim}{\bar{y}})$ in $\mathbb{Z}$ for which $\bar{w}(\underset{\sim}{h}) \not\geq e$, and thus $0\bar{w}(\underset{\sim}{h}) < 0$. We shall produce a substitution $z \to g(z)$ for $w(\underset{\sim}{x},\underset{\sim}{y})$ in $\mathbb{Z}\mathrm{Wr}\mathbb{Z}$ for which $w(\underset{\sim}{g}) \not\geq e$.

We claim that there is an integer vector $\underset{\sim}{t}$ such that

(a) $\underset{\sim}{t}$ is a solution of (3),

and (b) For *each* pair (i_1, k_1), (i_2, k_2), with $i_1, i_2 \in I_0$, such that $x_{i_1k_1}$ and $x_{i_2k_2}$ are the same variable z and (3) does not imply $D_{i_1k_1} = D_{i_2k_2}$ (i.e., such that these two occurrences of z were split in the formation of $\bar{w}$), we have

(5) $D_{i_1k_1}(\underset{\sim}{t}) \neq D_{i_2k_2}(\underset{\sim}{t})$.

Because of how the variables were split to form $\bar{w}$, this can be arranged for any *one* pair (i_1,k_1), (i_2,k_2). Among all solutions $\underset{\sim}{t}$ of (3), pick one which satisfies as many as possible of the inequalities (5). We claim that $\underset{\sim}{t}$ satisfies all the inequalities. If not, pick one not satisfied by $\underset{\sim}{t}$, so that

(6) $D_{i_1k_1}(\underset{\sim}{t}) = D_{i_2k_2}(\underset{\sim}{t})$.

Now pick a solution $\underset{\sim}{u}$ of (3) such that (6) fails for $\underset{\sim}{u}$. Then there exists an integer n sufficiently large that $\underset{\sim}{u} + n\underset{\sim}{t}$ not only satisfies (3) and satisfies the inequality which $\underset{\sim}{t}$ failed to satisfy, but also satisfies all the inequalities (5) which $\underset{\sim}{t}$ did satisfy. This violation of the way in which $\underset{\sim}{t}$ was picked shows that $\underset{\sim}{t}$ accomplishes (b) as well as (a).

Now we define the substitution $z \to g(z)$ for $w(\underset{\sim}{x},\underset{\sim}{y})$ in $\mathbb{Z}\mathrm{Wr}\mathbb{Z}$. Let $\hat{g}(z)$ be that coordinate of t corresponding to z. For each (i,k) such that $i \in I_0$, let $g(x_{ik})_m = h(\bar{z})$, where $m = D_{ik}(\underset{\sim}{t})$ and $\bar{z}$ is the variable in $\bar{w}(\underset{\sim}{\bar{x}},\underset{\sim}{\bar{y}})$ which replaced x_{ik} in w when the variables were split. Since $\underset{\sim}{t}$ satisfies condition (b), these components $g(x_{ik})_m$ are well defined. All local components $g(x_{ik})_m$ not defined above we take to be zero.

Suppose z is constrained, i.e., $z \in \underset{\sim}{x}$. If $z \in \underset{\sim}{x} \setminus \underset{\sim}{x}_0$, $\hat{g}(z) > 0$. If $z \in \underset{\sim}{x}_0$, then $\hat{g}(z) = 0$, and for each m either $g(z)_m = h(\bar{z}) \geq 0$ because $\bar{z}$ is constrained in w, or $g(z)_m = 0$. In either case, $e \leq g(z) \in \mathbb{Z}\mathrm{Wr}\mathbb{Z}$. Thus $z \to g(z)$ is a substitution.

Since t satisfies condition (a), $(0,0)w(\underset{\sim}{g}) = (0,0)\tilde{w}(g)$ $= (0\bar{w}(\underset{\sim}{h}), 0) < (0,0)$. Therefore our substitution shows that "$\forall \underset{\sim}{x} \geq \underset{\sim}{e}$, $\forall \underset{\sim}{y}$, $w(\underset{\sim}{x},\underset{\sim}{y}) \geq e$" fails in $\mathbb{Z}\mathrm{Wr}\mathbb{Z}$.

The preceding proof works just as well when $\mathbb{Z}\mathrm{Wr}\mathbb{Z}$ is changed to $G\mathrm{Wr}\mathbb{Z}$, where G is any ℓ-group. (In the first half of the proof, conjugation by an appropriate element assures that *some* point $(\beta,0) \in \mathbb{Z}_0$ is moved strictly down by $w(\underset{\sim}{g})$, and that suffices.) Moreover, letting $\mathcal{V}$ be the variety ℓ-var (G) of ℓ-groups generated by G, it turns out [3, Theorems 4.2 and 4.4] that the product variety $\mathcal{V}\mathcal{A}$ is generated by $G\mathrm{Wr}\mathbb{Z}$. Thus we have the following

THEOREM 7. Let $w = \bigvee_i \bigwedge_j w_{ij}$. Let G be any ℓ-group, and let $\mathcal{V} = \ell$-var (G). Then "$\forall \underset{\sim}{x} \geq \underset{\sim}{e}$, $\forall \underset{\sim}{y}$, $w(\underset{\sim}{x},\underset{\sim}{y}) \geq e$" holds in $\mathcal{V}\mathcal{A}$ if and only if for each $\dot{w} \in \mathcal{S}(w)$, "$\forall \underset{\sim}{\dot{x}} \geq \underset{\sim}{e}$, $\forall \underset{\sim}{\dot{y}}$, $\dot{w}(\underset{\sim}{\dot{x}},\underset{\sim}{\dot{y}}) \geq e$" holds in $\mathbb{Z}$ and "$\forall \underset{\sim}{\bar{x}} \geq \underset{\sim}{e}$, $\forall \underset{\sim}{\bar{y}}$, $\bar{w}(\underset{\sim}{\bar{x}},\underset{\sim}{\bar{y}}) \geq e$" holds in G. Thus if all sentences "$\forall \underset{\sim}{x} \geq \underset{\sim}{e}$, $\forall \underset{\sim}{y}$, $w(\underset{\sim}{x},\underset{\sim}{y}) \geq e$" are decidable in G, the same is true in the variety $\mathcal{V}\mathcal{A}$.

By induction, this theorem gives a decision procedure for the variety $\mathcal{A}^n$. To facilitate the statement of this procedure, we introduce some notation.

Let $\dot{w} \in \mathcal{S}(w)$. We denote $\dot{w}(\underset{\sim}{\dot{x}},\underset{\sim}{\dot{y}})$ by $\dot{w}_1(\underset{\sim}{x}_1,\underset{\sim}{y}_1)$. If "$\forall \underset{\sim}{x}_1 \geq \underset{\sim}{e}$, $\forall \underset{\sim}{y}_1$, $\dot{w}_1(\underset{\sim}{x}_1,\underset{\sim}{y}_1) \geq e$" holds in $\mathbb{Z}$, so that $\bar{w}(\underset{\sim}{\bar{x}},\underset{\sim}{\bar{y}})$ is defined, we denote the latter by $\dot{w}_2(\underset{\sim}{x}_2,\underset{\sim}{y}_2)$. In the same way, we use $\dot{w}_2$ to construct $\dot{w}_3(\underset{\sim}{x}_3,\underset{\sim}{y}_3)$, and so on, as long as the relevant sentences hold in $\mathbb{Z}$. If some sentence "$\forall \underset{\sim}{x}_h \geq \underset{\sim}{e}$, $\forall \underset{\sim}{y}_h$, $\dot{w}_h(\underset{\sim}{x}_h,\underset{\sim}{y}_h) \geq e$" fails in $\mathbb{Z}$, the process for this $\dot{w}$ terminates (at the hth stage), and "$\forall \underset{\sim}{x} \geq \underset{\sim}{e}$, $\forall \underset{\sim}{y}$, $\dot{w}(\underset{\sim}{x},\underset{\sim}{y}) \geq e$" holds in $\mathcal{A}^{h-1}$ but fails in $\mathcal{A}^h$. On the other hand, if some $\dot{w}_{h+1}$ is formally identical to $\dot{w}_h$, then again this process terminates (at the hth stage) -- all further $\dot{w}_m$'s are formally identical to $\dot{w}_h$, and "$\forall \underset{\sim}{x} \geq \underset{\sim}{e}$, $\forall \underset{\sim}{y}$, $\dot{w}(\underset{\sim}{\dot{x}},\underset{\sim}{\dot{y}}) \geq e$" holds in all $\mathcal{A}^n$'s. Of course, this affirmative termination is automatic if the system (2) for $\dot{w}_h$ has only the trivial solution.

ALGORITHM 8. (for $\mathcal{A}^n$). Let $w = \bigvee_i \bigwedge_j w_{ij}$. Then "$\forall \underset{\sim}{x} \geq \underset{\sim}{e}$, $\forall \underset{\sim}{y}$, $w(\underset{\sim}{x},\underset{\sim}{y}) \geq e$" fails in $\mathcal{A}^n$ if and only if for some $\dot{w} \in \mathcal{S}(w)$, and for some $1 \leq h \leq n$, "$\forall \underset{\sim}{x}_h \geq \underset{\sim}{e}$, $\forall \underset{\sim}{y}_h$, $\dot{w}_h(\underset{\sim}{x}_h,\underset{\sim}{y}_h) \geq e$" fails in $\mathbb{Z}$.

To test w in the variety $\mathcal{N}$ of normal valued ℓ-groups, we use the fact that "$\forall \underset{\sim}{x} \geq \underset{\sim}{e}$, $\forall \underset{\sim}{y}$, $w(\underset{\sim}{x},\underset{\sim}{y}) \geq e$" fails in $\mathcal{N}$ if and only if it fails in some $\mathcal{A}^n$ (as mentioned in the introduction). Thus for a given $\dot{w} \in \mathcal{S}(w)$, we process the sentences "$\forall \underset{\sim}{x}_h \geq \underset{\sim}{e}$, $\forall \underset{\sim}{y}_h$, $\dot{w}_h(\underset{\sim}{x}_h,\underset{\sim}{y}_h) \geq e$" until we reach termination. Each stage in this procedure must reduce either the number of group words, or the number of constrained variables, or the number of repeated occurrences of variables (i.e., the first occurrence of each variable is not counted). This gives a bound $B(\dot{w})$ for the number of stages involved, and establishes

ALGORITHM 9. (for $\mathcal{N}$). Let $w = \bigvee_i \bigwedge_j w_{ij}$. Then "$\forall \underset{\sim}{x} \geq \underset{\sim}{e}$, $\forall \underset{\sim}{y}$, $w(\underset{\sim}{x},\underset{\sim}{y}) \geq e$" fails in $\mathcal{N}$ if and only if for some $\dot{w} \in \mathcal{S}(w)$ and some $1 \leq h \leq B(\dot{w})$, "$\forall \underset{\sim}{x}_h \geq \underset{\sim}{e}$, $\forall \underset{\sim}{y}_h$, $\dot{w}_h(\underset{\sim}{x}_h,\underset{\sim}{y}_h) \geq e$" fails in $\mathbb{Z}$.

The solvability of the word problem in free normal valued ℓ-groups follows (modulo §5) from Algorithm 9. In fact the algorithm runs in non-deterministic polynomial time, though we shall not prove this. For the record, we state

THEOREM 10. The word problem is solvable in free normal valued ℓ-groups, and in ℓ-groups which are free in the variety $\mathcal{A}^n$.

EXAMPLE 11. Here we consider an example from "real life". In [3], a defining law for $\mathcal{A}^2$ is given, which after a very slight modification, reads:

$$\forall u, x, y, z \geq e,\ (u \wedge x^{-1}y^{-1}xy)(z \wedge x^{-1}y^{-1}xy) = (z \wedge x^{-1}y^{-1}xy)(u \wedge x^{-1}y^{-1}xy).$$

(In this form, though presumably not in the standard form below, the reader may be able to tell by inspection that this law holds in $\mathbb{Z}\text{wr}\mathbb{Z}$ but not in $\mathbb{Z}\text{wr}\mathbb{Z}\text{wr}\mathbb{Z}$. But here we want to illustrate our algorithms. One way of putting this law into standard form yields

(7) $\forall u, x, y, z \geq e,\ w(u,x,y,z) = e,$

where, letting $c = x^{-1}y^{-1}xy$, w is given by

$$\begin{aligned} w(u,x,y,z) = \\ &(u^{-1}z^{-1}uz \wedge u^{-1}z^{-1}uc \wedge u^{-1}z^{-1}cz \wedge u^{-1}z^{-1}cc) \\ \vee\ &(u^{-1}c^{-1}uz \wedge u^{-1}c^{-1}uc \wedge u^{-1}z \wedge u^{-1}c) \\ \vee\ &(c^{-1}z^{-1}uz \wedge c^{-1}z^{-1}uc \wedge c^{-1}z^{-1}cz \wedge c^{-1}z^{-1}c^{-1}c^{-1}) \\ \vee\ &(c^{-1}c^{-1}uz \wedge c^{-1}c^{-1}uc \wedge c^{-1}z \wedge e). \end{aligned}$$

The reader should envision this with each c actually replaced by $x^{-1}y^{-1}xy$.

At the moment we decide instead whether

(8) $\forall u, x, y, z, w \geq e$

holds in $\mathcal{N}$. We treat the "first" $\dot{w}$, i.e.,

$$\dot{w} = u^{-1}z^{-1}uz \vee u^{-1}c^{-1}uz \vee c^{-1}z^{-1}uz \vee c^{-1}c^{-1}uz.$$

The abelianizations 0, z, u, u+z are particularly simple because of the commutators, which are often present in naturally occurring words. The solutions of (2) are u = z = 0, with x and y arbitrary non-negative integers. Thus $\tilde{w} = \dot{w}$, and

$$w_2 = \bar{w} = u^{-1}z^{-1}uz \vee u^{-1}\hat{c}^{-1}uz \vee \hat{c}^{-1}z^{-1}uz \vee \hat{c}^{-1}\hat{c}^{-1}uz,$$

where $\hat{c} = x_1^{-1}y_1^{-1}x_2y_2$, with x_1, x_2, y_1, y_2 unconstrained. This time the abelianizations are 0, $x_1 + y_1 - x_2 - y_2 + z$, $x_1 + y_1 - x_2 - y_2 + u$, $2x_1 + 2y_1 - 2x_2 - 2y_2 + u + z$. We get $\tilde{w}_2 = u^{-1}z^{-1}uz$, and since u and z are not forced by (2) to be zero, we get $w_3 = \bar{w}_2 = u_1^{-1}z_1^{-1}u_2z_2$, which can obviously be negated in $\mathbb{Z}$. Thus for this $\dot{w}$, $\dot{w} \geq e$ holds in $\mathbb{Z}\mathrm{Wr}\mathbb{Z}$ but not in $\mathbb{Z}\mathrm{Wr}\mathbb{Z}\mathrm{Wr}\mathbb{Z}$. This shows (in three stages) that (8) (and thus also (7)) fails in $\mathcal{N}$ and indeed in $\mathcal{A}^3$, though it does not by itself show that (8) holds in $\mathcal{A}^2$.

What if this "$\dot{w} \geq e$" had held in $\mathcal{N}$? We would have had to proceed through the rest of the $4^3 \cdot 3$ $\dot{w}$'s until we found one which failed in $\mathcal{N}$, or found that all held in $\mathcal{N}$. Later, with the machinery of §4, we shall see that if (7) had held in $\mathcal{N}$, this would not after all have been so hard to ascertain.

If Algorithm 9 shows that "$\forall \underset{\sim}{x} \geq \underset{\sim}{e}, \forall \underset{\sim}{y}, w(\underset{\sim}{x},\underset{\sim}{y}) \geq e$" fails in $\mathcal{N}$, then the data generated during the execution of the algorithm can be used to write down a specific substitution in $\mathrm{Wr}^n\mathbb{Z}$ demonstrating this failure, where n is the number of stages which was required for the $\dot{w}$ for which the failure occurred; provided, of course, that the method of determining the existence of integer solutions of the various systems allows one to write down specific solutions. This proviso holds if one uses the Motzkin Algorithm of §5. Thus when a negative answer is obtained, one may verify that no error has been made in the execution of the algorithm without checking all of the work done during the execution, or may demonstrate the correctness of the answer to someone unfamiliar with the algorithm. Unfortunately, no such convenient device is available when the answer is positive.

EXAMPLE 12. With the present machinery, it is easy to establish that the defining laws for $\mathcal{N}$ mentioned in the introduction do indeed hold in $\mathcal{N}$. We consider

$$\forall x,\ z \geq e,\ (x^2 z^{-2} \wedge x^{-2} z^2)^2 \leq x \wedge z.$$

One way of putting this law into standard form yields

(9) $\forall x,\ z \geq e,\ w(x,z) \geq e,$

where $w(x,z) =$

$$\begin{aligned} &(z^2 x^{-2} z^2 x^{-1} \wedge z^2 x^{-2} z^2 x^{-2} z) \\ \vee\ &(z^{-2} x^2 z^2 x^{-1} \wedge z^{-2} x^2 z^2 x^{-2} z) \\ \vee\ &(z^2 x^{-2} z^{-2} x^3 \wedge z^2 x^{-2} z^{-2} x^2 z) \\ \vee\ &(z^{-2} x^2 z^{-2} x^3 \wedge z^{-2} x^2 z^{-2} x^2 z) \end{aligned}$$

The "matrix" of abelianizations $C(w_{ij})$ is

$$\begin{matrix} -3x + 4z & -4x + 5z \\ x & z \\ x & z \\ 5x - 4z & 4x - 3z \end{matrix}$$

Since any $\dot{w}$ must include a w_{ij} whose abelianization is in the second row of the matrix, and since x and z are constrained to be non-negative, the only solution of the system (2) is the trivial solution, so the algorithm terminates in one stage with a positive answer--for any $\dot{w}$ whatever. Therefore (9) holds in $\mathcal{N}$.

More generally, when dealing with " $\forall x,\ z \geq e,\ (x^2 z^{-2} \wedge x^{-2} z^2)^n \leq x \wedge z$, the matrix of abelianizations contains the row

$$\begin{matrix} x & z \end{matrix}$$

if n is even, and contains the rows

$$\begin{array}{ll} -x + 2z & -2x + 3z \\ 3x - 2z & 2x - z \end{array}$$

if n is odd, so that for any $\dot{w}$ the system (2) has only the trivial solution, and the law holds in $\mathcal{N}$. The other laws in the introduction are even easier, with the algorithm terminating in the first stage in all cases.

EXAMPLE 13. Surely it is appropriate to offer here one new defining law for $\mathcal{N}$:

(10) $\forall\, x,\ z \geq e,\ x^2z^{-1} \vee x^{-1}z \geq e.$

The abelianizations are $2x - z$ and $-x + z$, and it is easily checked by drawing graphs that (2) has only the trivial solution, so that (10) holds in $\mathcal{N}$. Also, it is easy to see that (10) fails in some ℓ-groups, e.g., in the ℓ-group $A(\mathbb{R})$ of all order-automorphisms of the real line $\mathbb{R}$. Therefore (10) is a defining law for $\mathcal{N}$. A similar argument shows that more generally, for $p,q,r,s > 0$,

(11) $\forall\, x,\ z \geq e,\ x^pz^{-q} \vee x^{-r}z^s,$

is a defining law for $\mathcal{N}$ provided that $p/q > r/s$.

4. THE STATEMENTS "$\forall\, \underset{\sim}{x} \geq \underset{\sim}{e},\ \forall\, \underset{\sim}{y},\ w(\underset{\sim}{x},\underset{\sim}{y}) \leq e$" and "$\forall\, \underset{\sim}{x} \geq \underset{\sim}{e},\ \forall\, \underset{\sim}{y},\ w(\underset{\sim}{x},\underset{\sim}{y}) = e$".

Let $w = \bigvee_i \bigwedge_j w_{ij}$. Obviously the above statements can be decided by putting w^{-1} in standard form and applying the previous techniques to decide whether $\forall\, \underset{\sim}{x} \geq \underset{\sim}{e},\ \forall\, \underset{\sim}{y},\ w^{-1}(\underset{\sim}{x},\underset{\sim}{y}) \geq e$. However, this would be extremely inefficient. The following algorithm requires testing only $|I|$ subwords in a way analogous to the testing of a single $\dot{w}$ in §3, and thus will ordinarily require much less effort than Algorithm 9.

ALGORITHM 14. Let $w = \bigvee_{i \in I} \bigwedge_{j \in J_i} w_{ij}$. "$\forall\, \underset{\sim}{x} \geq \underset{\sim}{e},\ \forall\, \underset{\sim}{y},\ w(\underset{\sim}{x},\underset{\sim}{y}) \leq e$" holds in $\mathcal{N}$ if and only if for each $i \in I$ (with no $w_{ij} = e$), "$\forall\, \underset{\sim}{x} \geq \underset{\sim}{e},\ \forall\, \underset{\sim}{y},\ \bigwedge_{j \in J_i} w_{ij} \leq e$" holds in $\mathcal{N}$. The latter statements are decided exactly as in §3, except that all inequalities except "$x \geq 0$" are reversed.

An analogous algorithm holds for $\mathscr{A}^n$. To decide whether "$\forall\, \underset{\sim}{x} \geq \underset{\sim}{e}$, $\forall\, \underset{\sim}{y}$, $w(\underset{\sim}{x},\underset{\sim}{y}) = e$", it obviously suffices to decide the two statements with "=" replaced by "$\geq$" and "$\leq$", respectively. But for $\mathscr{N}$, we shall do better; see Algorithm 19.

LEMMA 15. Let $w = \bigvee_i \bigwedge_j w_{ij}$. Suppose "$\forall\, \underset{\sim}{x} \geq \underset{\sim}{e}$, $\forall\, \underset{\sim}{y}$, $w(\underset{\sim}{x},\underset{\sim}{y}) \geq e$" (resp., $\leq e$) fails in $\mathscr{N}$. Let $z \to g(z)$ be a substitution in some wreath power $\mathrm{Wr}^n\mathbb{Z}$ for which $\underset{\sim}{0}w_{ij}(\underset{\sim}{g}) = \underset{\sim}{0}$ for each w_{ij}. Then there exists another wreath power $\mathrm{Wr}^m\mathbb{Z}$ such that there is a substitution $z \to h(z)$ in $\mathrm{Wr}^{m+n}\mathbb{Z}$ which matches the given substitution $z \to g(z)$ on the top n levels and for which $w(\underset{\sim}{h}) \not\geq e$ (resp., $\not\leq e$). (Intuitively, prescribing what each variable shall do on several wreath layers, in such a way that $\underset{\sim}{0}w_{ij} = \underset{\sim}{0}$ for each w_{ij}, cannot prevent demonstration of the failure of $w \geq e$ or of $w \leq e$ in a sufficiently much larger wreath power.)

Proof. We treat the case "$w \geq e$". Since "$w \geq e$" fails in $\mathscr{N}$, it fails in some $\mathrm{Wr}^m\mathbb{Z}$. Let $z \to k(z)$ be a substitution in $\mathrm{Wr}^m\mathbb{Z}$ demonstrating this failure. In $\mathrm{Wr}^{m+n}\mathbb{Z}$, each point r in the set on which $\mathrm{Wr}^n\mathbb{Z}$ acts is replaced by a copy S_r of the set S on which $\mathrm{Wr}^m\mathbb{Z}$ acts. We define $z \to h(z)$ so as to match $z \to g(z)$ on the top n levels of $\mathrm{Wr}^{m+n}\mathbb{Z}$, and so that the action of each $h(z)$ on *each* S_r matches that of $k(z)$ even as S_r is moved to a different $S_{r'}$. Since $\underset{\sim}{0}w_{ij}(\underset{\sim}{g}) = 0$ for each w_{ij}, $S_{\underset{\sim}{0}}w_{ij}(\underset{\sim}{k}) = S_{\underset{\sim}{0}}$ for each w_{ij}. Then since each variable acts the same way on all the S_r's, the point $\underset{\sim}{0}$ of Wr^{m+n} is moved down by $w(\underset{\sim}{k})$.

THEOREM 16. Let $w = \bigvee_i \bigwedge_j w_{ij}$, with no $w_{ij} = e$, and suppose that "$\forall\, \underset{\sim}{x} \geq \underset{\sim}{e}$, $\forall\, \underset{\sim}{y}$, $w(\underset{\sim}{x},\underset{\sim}{y}) \geq e$" holds in $\mathscr{N}$. Let u be any ℓ-group word. Then "$\forall\, \underset{\sim}{x} \geq \underset{\sim}{e}$, $\forall\, \underset{\sim}{y}$, $w(\underset{\sim}{x},\underset{\sim}{y}) \wedge u(\underset{\sim}{x},\underset{\sim}{y}) \leq e$" holds in $\mathscr{N}$ only if "$\forall\, \underset{\sim}{x} \geq \underset{\sim}{e}$, $\forall\, \underset{\sim}{y}$, $u(\underset{\sim}{x},\underset{\sim}{y}) \leq e$" holds in $\mathscr{N}$.

Proof. Suppose "$\forall\, \underset{\sim}{x} \geq \underset{\sim}{e}$, $\forall\, \underset{\sim}{y}$, $w \wedge u \leq e$" holds in $\mathscr{N}$ but "$\forall\, \underset{\sim}{x} \geq \underset{\sim}{e}$, $\forall\, \underset{\sim}{y}$, $u \leq e$" fails. Then there exists a substitution $z \to g(z)$ in some $\mathrm{Wr}^n\mathbb{Z}$ for which $\underset{\sim}{0} < \underset{\sim}{0}u(\underset{\sim}{g})$. (By the transitivity of $\mathrm{Wr}^n\mathbb{Z}$, we may assume that this holds for the particular point $\underset{\sim}{0}$.) Since $w \wedge u \leq e$, $\underset{\sim}{0} = \underset{\sim}{0}w(\underset{\sim}{g}) = \max_i \min_j \underset{\sim}{0}w_{ij}(\underset{\sim}{g})$. Let $I' = \{i \in I \mid \underset{\sim}{0} = \min_j \underset{\sim}{0}w_{ij}(\underset{\sim}{g})\} \neq \phi$. For each $i \in I'$, let $J_i' = \{j \in J_i \mid \underset{\sim}{0} = \underset{\sim}{0}w_{ij}(\underset{\sim}{g})\} \neq \phi$. Let $w' = \bigvee_{i \in I'} \bigwedge_{j \in J_i'} w_{ij}$.

Then "$\forall \tilde{x} \geq \tilde{e}$, $\forall \tilde{y}$, $w' \geq e$" holds in $\mathcal{N}$, for if not, by applying Lemma 15 to w', we would obtain a substitution in some $\mathrm{Wr}^{m+n}\mathbb{Z}$ which agreed with $z \to g(z)$ on the top n levels and in which $\tilde{0}$ would be moved down by w' and thus also by w (because of the way w' was constructed), violating "$w \geq e$". On the other hand, $w' \leq e$ in $\mathcal{N}$, for if not, by another application of Lemma 15 to w', we would obtain a substitution in some $\mathrm{Wr}^{m+n}\mathbb{Z}$ in which $\tilde{0}$ would be moved up by w' and also by u since $\tilde{0} < \tilde{0}u(\tilde{g})$, violating $w \wedge u \leq e$.

Therefore, "$\forall \tilde{x} \geq \tilde{e}$, $\forall \tilde{y}$, $w = e$" holds in $\mathcal{N}$ even though no $w_{ij} = e$, violating Lemma 1.

COROLLARY 17. Let N be the free normal valued ℓ-group on $\tilde{y}$. Let $w = \bigvee_i \bigwedge_j w_{ij}$, with the w_{ij}'s group words in $\tilde{y}$ and each $w_{ij} \neq e$. Then $|w(\tilde{y})|$ is a weak order unit of N.

Proof. If no group word of w is e, then $|w|$ can be put in standard form so that no group word of $|w|$ is e.

Corollary 17 cannot be made to say that $w(\tilde{x} \vee \tilde{e}, \tilde{y})$ is a weak order unit in the free normal valued ℓ-group on $\tilde{x} \cup \tilde{y}$. For example, let $w(\tilde{x},\tilde{y}) = x$, a single constrained variable. Then $w(\tilde{x} \vee \tilde{e}, \tilde{y}) = x \vee e$, and of course $(x \vee e) \wedge (x^{-1} \vee e) = e$. This does not violate Theorem 16 because $x^{-1} \vee e$ is not an ℓ-group word in $x \vee e$.

COROLLARY 18. Let $w = \bigvee_i w_i = \bigvee_i \bigwedge_j w_{ij}$. When deciding whether "$\forall \tilde{x} \geq \tilde{e}$, $\forall \tilde{y}$, $w(\tilde{x},\tilde{y}) \geq e$" or "$\forall \tilde{x} \geq \tilde{e}$, $\forall \tilde{y}$, $w(\tilde{x},\tilde{y}) = e$" holds in $\mathcal{N}$, any term w_i which contains no e and for which "$\forall \tilde{x} \geq \tilde{e}$, $\forall \tilde{y}$, $w_i(\tilde{w},\tilde{y}) \leq e$" holds in $\mathcal{N}$ may be erased before the testing is begun. If all terms are thus erased, the statements fail.

Proof. This corollary follows from Theorem 16 and Lemma 1. To see that the first statement fails if all terms are erased, observe that certainly all but one term w_i can be erased, and then since "$\forall \tilde{x} \geq \tilde{e}$, $\forall \tilde{y}$, $w_i(\tilde{x},\tilde{y}) \leq e$" holds in $\mathcal{N}$, all group words but one w_{ij} may be erased. Now for the original statement to hold, "$\forall \tilde{x} \geq \tilde{e}$, $\forall \tilde{y}$, $w_{ij} = e$" would have to hold, violating Lemma 1.

ALGORITHM 19. (for deciding whether "$\forall \tilde{x} \geq \tilde{e}$, $\forall \tilde{y}$, $w(\tilde{x},\tilde{y}) = e$" holds in $\mathcal{N}$). Let $w = \bigvee_i w_i = \bigvee_i \bigwedge_j w_{ij}$.

(a) Test "$\forall \underset{\sim}{x} \geq \underset{\sim}{e}$, $\forall \underset{\sim}{y}$, $w(\underset{\sim}{x},\underset{\sim}{y}) \leq e$" in $\mathcal{N}$, using Algorithm 14. If it fails, the answer to the original question is negative; if it holds, go to (b).

(b) Form w' be deleting from w each term w_i which contains no e. If all terms are deleted, the answer is negative. If not, the answer is the same as for "$\forall \underset{\sim}{x} \geq \underset{\sim}{e}$, $\forall \underset{\sim}{y}$, $w'(\underset{\sim}{x},\underset{\sim}{y}) \geq e$", which can be tested by Algorithm 9.

Proof. Suppose "$\forall \underset{\sim}{x} \geq \underset{\sim}{e}$, $\forall \underset{\sim}{y}$, $w(\underset{\sim}{x},\underset{\sim}{y}) \leq e$" holds in $\mathcal{N}$. Then for each w_i, "$\forall \underset{\sim}{x} \geq \underset{\sim}{e}$, $\forall \underset{\sim}{y}$, $w_i(\underset{\sim}{x},\underset{\sim}{y}) \leq e$" holds in $\mathcal{N}$, and we can apply Corollary 18.

THEOREM 20. Results 16, 17, and 18 also hold for the varieties $\mathcal{L}$ (of all ℓ-groups) and $\mathcal{A}$, where for $\mathcal{A}$ the assumption about the group words w_{ij} is that no abelianized w_{ij} is 0.

Proof. We shall comment only on the analog of Theorem 16. The proof for $\mathcal{L}$ is like the proof for $\mathcal{N}$, and involves a wreath product of two (not necessarily normal valued) ℓ-permutation groups.

If $u \leq e$ fails in $\mathcal{A}$, then there exists a substitution $z \to g(z)$ in the additive rationals Q for which $u(\underset{\sim}{g}) > 0$. We have $w(\underset{\sim}{g}) = 0$ only on a subset of the union of a finite number of hyperplanes in d(w)-dimensional rational space, one for each w_{ij}. (Since no abelianized w_{ij} is 0, each $C(w_{ij})$ has a non-zero coefficient.) Proceding one at a time through the w_{ij}'s and using continuity, we can modify the g(z)'s slightly, so that $w(\underset{\sim}{g}) \neq 0$ (and thus $w(\underset{\sim}{g}) > 0$) and still $u(\underset{\sim}{g}) > 0$. This shows that $w \wedge u \leq e$ fails in $\mathcal{A}$. (Alternately, the abelian case can be deduced from results in [10].)

5. THE MOTZKIN ALGORITHM FOR SYSTEMS OF INEQUALITIES

Given a word $\dot{w}(\underset{\sim}{x},\underset{\sim}{y}) = \bigvee_{i \in I} w_i$, we must be able to determine whether the system

$$(1) \quad \begin{array}{ll} x \geq 0 & (x \in \underset{\sim}{x}), \\ C_i < 0 & (i \in I) \end{array}$$

has a solution in integers (equivalently in rationals); and if not, then for the system

$$(2) \quad \begin{array}{ll} x \geq 0 & (x \in \underset{\sim}{x}), \\ C_i \leq 0 & (i \in I), \end{array}$$

we must be able to find $\underset{\sim}{x}_0 = \{x \in \underset{\sim}{x} \mid (2) \text{ implies } x = 0\}$ and $I_0 = \{i \in I \mid (2)$ implies $C_i = 0\}$.

If there is a C_i whose coefficients are all zero, (1) cannot have a solution, and all such i's are automatically included in I_0, permitting us to ignore them in dealing with (2). Letting $I_{00} = \{i \in I \mid$ all coefficient of C_i are $0\}$, we treat

$$(2') \quad \begin{array}{ll} x \geq 0 & (x \in \underset{\sim}{x}), \\ C_i \leq 0 & (i \in I \setminus I_{00}). \end{array}$$

Each inequality in (2') represents a half space.

There is an algorithm known as the double description method which is superbly suited to our purposes. This very efficient algorithm (which runs in polynomial time) was created by T. S. Motzkin, and is described in [9]. The Motzkin Algorithm treats systems of non-strict linear inequalities $D \geq 0$ in d unknowns. (Here $D = -C_i$ and $d = |\underset{\sim}{x} \cup \underset{\sim}{y}|$.) Requiring $x \geq 0$ for certain unknowns necessitates virtually no extra effort. The algorithm produces a finite (possibly empty) set $A \cup B$ of vectors in d-dimensional rational space ("rational" provided the coefficients are rational, and here they are integers). *The rational solutions of the system are precisely the linear combinations of the vectors in* $A \cup B$ *using non-negative rational scalars,* and $A \cup B$ is a minimal set of vectors having this property. *Each* $\underset{\sim}{a} \in A$ *makes each* $C_i(\underset{\sim}{a}) = 0$ *and makes* $x = 0$ *for each constrained variable* x*; and no* $\underset{\sim}{b} \in B$ *does this.* As an aid to geometric visualization, we mention that if A is empty, B consists of the "edges" of a polyhedral cone in d-space, with the cone perhaps having dimension smaller than d.

ALGORITHM 21 (for deciding whether the system (1) has a solution, and for finding I_0 and $\underset{\sim}{x}_0$). Let $\dot{w}(\underset{\sim}{x},\underset{\sim}{y}) = \bigvee_{i \in I} w_i$. Apply the Motzkin Algorithm to (2'). Then $I_0 = I_{00} \cup \{i \in I \mid C_i(\underset{\sim}{b}) = 0 \text{ for all } \underset{\sim}{b} \in B\}$, and $\underset{\sim}{x}_0 = \{x \in \underset{\sim}{x} \mid x = 0 \text{ for all } \underset{\sim}{b} \in B\}$. Moreover, (1) has a solution if and only if I_0 is empty.

Proof. This result follows from the above description of solutions, and of A.

ALGORITHM 22. In Algorithm 5, the set A provided here by the Motzkin Algorithm may be used in lieu of F.

Proof. A is a spanning set for the solution space W of system (4), as can be seen from the above description of solutions, and of A and B.

Although A is sometimes larger than F, if one is using the Motzkin Algorithm, then A is already known prior to splitting, whereas F has yet to be computed.

6. ADVANTAGES OF BEING CLEVER

When making decisions in $\mathcal{N}$ (but not in $\mathcal{A}^n$) by using Algorithm 9 or 14 it is often possible to save effort in the processing of a given $\dot{w}$ in the following non-algorithmic way. At any stage in the execution, one may happen to spot a solution of system (2) which yields some progress (dropping of terms w_i, unconstraining of variables, or splitting of variables based on whether $D_{i_1k_1} = D_{i_2k_2}$ is implied by those equations $C_i = 0$ and $x = 0$ which are in fact satisfied by this solution of (2)). One may use this solution of (2) to form $\bar{w}$, and even if the progress made is less than could have been accomplished by a full execution of that stage, there is often a great deal of progress per amount of effort expanded. It does not matter if the existence of a solution of (1) is overlooked.

For example, if several variables are unconstrained, one may choose to execute a stage consisting solely of splitting the variables as in Algorithm 9, but with (4) replaced by

$$(4') \quad \begin{array}{l} x = 0 \ (x \in \underset{\sim}{x}), \\ C_i = 0 \ (i \in I). \end{array}$$

Here finding F requires only the solution of a homogeneous system of linear equations.

7. DECIDING DISJUNCTIVE FORMULAS

A *disjunctive formula* is a disjunction of conjunctions of atomic formulas. An equivalent disjunctive formula can be obtained by replacing each conjunction of atomic formulas by a single atomic formula. (Replace

"$u(\underset{\sim}{y}) = e$ & $w(\underset{\sim}{y}) = e$" by "$u(\underset{\sim}{y})w(\underset{\sim}{z}) = e$", where $\underset{\sim}{y}$ and $\underset{\sim}{z}$ are disjoint sets of variables.)

LEMMA 23. Let $P(\underset{\sim}{x},\underset{\sim}{y})$ be the predicate "$u_1 \geq e$ or $u_2 \geq e$ or...or $u_m \geq e$" (resp., $\leq e$, $= e$), where each $u_i = u_i(\underset{\sim}{x},\underset{\sim}{y})$ is an ℓ-group word. Then "$\forall \underset{\sim}{x} \geq \underset{\sim}{e}$, $\forall \underset{\sim}{y}$, $P(\underset{\sim}{x},\underset{\sim}{y}) \geq e$" (resp., $\leq e$, $= e$) holds in $\mathscr{N}$ if and only if for some $i = 1,\ldots,m$, "$\forall \underset{\sim}{x} \geq \underset{\sim}{e}$, $\forall \underset{\sim}{y}$, $u_i(\underset{\sim}{x},\underset{\sim}{y}) \geq e$" (resp., $\leq e$, $= e$) holds in $\mathscr{N}$.

Proof. Sufficiency being obvious, we prove necessity. We treat the case involving "$u_i \geq e$", the other cases being similar.

Thus suppose that for each $i = 1,\ldots,m$ "$\forall \underset{\sim}{x} \geq \underset{\sim}{e}$, $\forall \underset{\sim}{y}$, $u_i(\underset{\sim}{x},\underset{\sim}{y}) \geq e$" fails in $\mathscr{N}$ and thus in some $\mathrm{Wr}^{n_i}\mathbb{Z}$. Let $n = \max_i n_i$. Then it is easy to see that "$\forall \underset{\sim}{x} \geq \underset{\sim}{e}$, $\forall \underset{\sim}{y}$, $P(\underset{\sim}{x},\underset{\sim}{y})$" fails in $(\mathrm{Wr}^n\mathbb{Z})\mathrm{Wr}\mathbb{Z} = \mathrm{Wr}^{n+1}\mathbb{Z}$.

THEOREM 24. Disjunctive formulas are decidable in $\mathscr{N}$. Moreover, a disjunctive formula holds in $\mathscr{N}$ if and only if it holds in each wreath power $\mathrm{Wr}^n\mathbb{Z}$.

An analogous lemma for the variety $\mathscr{L}$ of all ℓ-groups can be proved by using the ℓ-group $A(\mathbb{R})$ of all order automorphisms of the real line $\mathbb{R}$, and and noting that any finite direct product of copies of $A(\mathbb{R})$ can be embedded in $A(\mathbb{R})$. By virtue of the procedure in [6] for deciding for an ℓ-group word u whether "$\forall \underset{\sim}{x} \geq \underset{\sim}{e}$, $\forall \underset{\sim}{y}$, $u(\underset{\sim}{x},\underset{\sim}{y}) \geq e$" (resp., $\leq e$, $= e$) holds in $\mathscr{L}$, we have

THEOREM 25. Disjunctive formulas are decidable in $\mathscr{L}$. Moreover, a disjunctive formula holds in $\mathscr{L}$ if and only if it holds in $A(\mathbb{R})$.

In fact, $A(\mathbb{R})$ can be replaced in Theorem 25 by any ℓ-group which is not normal valued.

REFERENCES

1. L. Fuchs, *Partially ordered algebraic systems*. Addison-Wesley, Reading, Mass., 1963.
2. A. M. W. Glass, *Ordered permutation groups*. Bowling Green State University, Bowling Green, Ohio, 1976.
3. A. M. W. Glass, W. C. Holland, and S. H. McCleary, The structure of ℓ-group varieties. *Algebra Universalis*. To appear.
4. W. C. Holland, The largest proper variety of lattice ordered groups. *Proc. Amer. Math. Soc.* 57: 25-28 (1976).

5. W. C. Holland and S. H. McCleary, Wreath products of ordered permutation groups. *Pacific J. Math.* 31: 703-716 (1969).

6. W. C. Holland and S. H. McCleary, Solvability of the word problem in free lattice-ordered groups. *Houston J. Math.* 5: 99-105 (1979).

7. N. G. Kisamov, Universal theory of lattice-ordered abelian groups. *Algebra and Logic* 5: 71-76 (1966).

8. S. H. McCleary, Generalized wreath products viewed as sets with valuation. *J. of Algebra* 16: 163-182 (1970).

9. T. S. Motzkin, H. Raiffa, G. L. Thompson, and R. M. Thrall, The double description method, *Contribution to the theory of games (vol. 2)*. H. W. Kuhn and A. W. Tucker (editors), Annals of Mathematical Studies, no. 28, Princeton University Press, Princeton, N. J., 1953.

10. E. Weinberg, Free lattice-ordered abelian groups. *Math. Ann.* 151: 187-199 (1963).

TRYING TO RECOGNIZE THE REAL LINE

W. Charles Holland
Bowling Green State University
Bowling Green, Ohio

It is possible to view any problem about lattice-ordered groups as a problem about the lattice-ordered group $A(\Omega)$ of order-preserving permutations of a chain Ω. Although this viewpoint has its detractors, there is by now a sizeable population and I count myself a member whose first inclination, when presented with a question about ℓ-groups, is to draw a picture. Must ℓ-groups have unique root extraction? No - draw a picture. Does the commutator equation $[[g^+,f^-],[g^-,f^+]] = e$ hold in every ℓ-group? Yes - draw a picture. Is the word problem for free ℓ-groups solvable? Yes - observe the choices you make while drawing a picture. The pictures that are actually drawn, or at least visualized, are representations of graphs of monotone increasing functions from a chain Ω to itself. In the typical problems it doesn't really matter what chain Ω is, as long as it is doubly homogeneous - to insure that any reasonable picture actually represents the graphs of permutations. In particular, I showed [1] that any equation which holds in one 0-2-transitive $A(\Omega)$ holds in all. George McNulty pointed out that, in fact, any positive universal statement which is true in one 0-2-transitive $A(\Omega)$ is true in all. This fact, together with experience, tempted me to conjecture that any elementary statement true in one 0-2-transitive $A(\Omega)$ is true in all. This paper is a report of my progress in discovering how wrong that conjecture was. In fact, I came very close to showing that the real line $\mathbb{R}$ is completely determined by elementary statements about $A(\mathbb{R})$. Soon after my presentation of these results at the Boise conference, the crucial final step was made by Yuri Gurevich. In a forthcoming paper by Gurevich and me [2], we give details of the proof that both $A(\mathbb{R})$ and $A(Q)$, Q the rational line, are elementarly defineable (among transitive $A(\Omega)$). In the present paper,

I will outline some of the facts leading up to that theorem which were known to me before the conference. Subsequent to the conference, and independently, there appeared the thesis of Michile Jambu-Girandet [3] which contains all of the results reported at the conference, and much more (but not the fact that $A(\mathbb{R})$ and $A(Q)$ are elementarily defined).

The elementary language for $A(\Omega)$ allows symbols for and, or, not, equals, variables, the group and lattice operations, and quantification over members of $A(\Omega)$ (but not over Ω or subsets of either Ω or $A(\Omega)$). The goal is to translate general statements about Ω or $A(\Omega)$ into equivalent elementary statements about $A(\Omega)$. In all that follows, I assume that $A(\Omega)$ is transitive.

A convex cycle $f \in A(\Omega)$ has the property that for any $\alpha \in \Omega$ which is moved by f, the set $\{\alpha f^n | n = 0, \pm 1, \pm 2, \ldots\}$ is coterminal in the support of f. Positive convex cycles f are characterized in $A(\Omega)$ by the elementary statement that f is not the join of two non-trivial disjoint positive elements. In $A(\mathbb{R})$, translation by +1 is a positive convex cycle which is disjoint from nothing. If $A(\Omega)$ has a positive convex cycle f which is disjoint from nothing, then the support of f is Ω, and Ω must be countably coterminal. This serves to distinquish $\mathbb{R}$ from such 0-2-homogeneous chains as the long line.

In $A(\mathbb{R})$, there are just four conjugacy classes of positive convex cycles: (1) those whose support is $\mathbb{R}$, (2) those whose support is a lower-bounded ray, (3) those whose support is an upper-bounded ray, and (4) those whose support is bounded. Because it is always true in permutation groups that support$(f^{-1}gf)$ is (support g)f, we see that in $A(Q)$, a convex cycle whose support is the interval $0 < x < 1$ cannot be conjugate to a convex cycle whose support is the interval $0 < x < \sqrt{2}$. Variations on this observation show that there are nine conjugacy classes of positive convex cycles in $A(Q)$. It is easy to express this in an elementary statement, and so $A(\mathbb{R})$ and $A(Q)$ are not elementarily equivalent.

Suppose now that $A(\Omega)$ is transitive and that $A(\Omega)$ is elementarily equivalent to $A(\mathbb{R})$. What can be said about Ω? By the earlier observation, we know that Ω must be countably coterminal. $A(\mathbb{R})$ contains no basic elements, and hence neither does $A(\Omega)$. This implies that every interval of Ω contains the support of some non-trivial member of $A(\Omega)$. We observe that in $A(\mathbb{R})$ or $A(\Omega)$, the fact that the support of f lies entirely below the support of g can be expressed by the elementary statement: $|f| \wedge h^{-1}|g|h = e$ for all positive h. Consider an element $f \in A(\mathbb{R})$ such that for some conjugate f' of f, the support of f is entirely below the support of f'. The support of f must then be bounded, and if α = glb(support f) and β = lub(support f), there exists $g \in A(\mathbb{R})$ such that $\alpha g = \beta$, so that $\beta = \text{glb(support } g^{-1}fg)$. There does not exist any permutations whose support is between the support of g and the support of $g^{-1}fg$. If Ω is not dedekind complete, let Σ denote a non-empty subset of Ω which is bounded above but has no least upper bound, and let $\alpha \in \Sigma$. Because every interval of Ω contains the support of a member of $A(\Omega)$, there is $f \in A(\Omega)$ such that $\alpha = \text{glb(support of } g)$ and support of f is cofinal in Σ. For every conjugate $g^{-1}fg$ whose support is above the support of f, $\Sigma < \alpha g = \text{glb(support of } g^{-1}fg)$, and as Σ has no least upper bound, the interval between Σ and αg contains the support of some member of $A(\Omega)$. The properties of this g have been described by elementary statements, and there is no such element of $A(\mathbb{R})$. It follows that Ω must be dedekind complete. A consequence of this is that $A(\Omega)$ is 0-2-transitive and every point of Ω has countable character.

It is possible to get a little more information about Ω with a lot more effort. Every positive permutation in $A(\Omega)$ or $A(\mathbb{R})$ is uniquely expressible as $f = \bigvee f_\alpha$ where each f_α is a convex cycle, and if $\alpha \neq \beta$, $f_\alpha \wedge f_\beta = e$. The f_α's are the convex cyclic factors of f. A join, g, of a subset of $\{f_\alpha\}$ is characterized by the condition $g \wedge fg^{-1} = e$. The set of convex cyclic factors of f has an induced total order by letting $f_\alpha < f_\beta$ mean that the support of f_α is below the support of f_β. Any permutation in the normalizer of f permutes the convex cyclic factors of f by conjugation, and preserves the induced order. If f is disjoint from nothing, the converse is also true: every order-preserving permutation of the set of convex cyclic factors of f is induced (under conjugation) by a member of the normalizer of f. Denote the ℓ-group of order-preserving permutations of the set of convex cyclic factors of f by $A(f)$. It is possible to translate any elementary statementsabout $A(f)$ into an elementary statement about $A(\Omega)$ or $A(\mathbb{R})$. In fact, by the preceeding remarks, we may even quantify over subsets of the set of convex cyclic factors of f. There are chains Ω, however, which are homogeneous, dedekind complete, and countably coterminal, and such that $A(\Omega)$ has an element f whose ordered set of convex cyclic factors is isomorphic to $\mathbb{R}$. By the preceeding discussion, $A(f)$ cannot be elementarily equivalent to $A(g)$ for any $g \in A(\mathbb{R})$, and so $A(\Omega)$ is not elementarily equivalent to $A(\mathbb{R})$.

These techniques are an attempt in the direction of capturing the Suslin property of $\mathbb{R}$ (every pair-wise disjoint collection of open sets is countable). In fact, a forthcoming paper of Glass, Gurevich, and Holland, uses a refinement of these techniques to characterize homogeneous continua satisfying the Suslin condition.

REFERENCES

1. W. C. Holland, The largest proper variety of lattice ordered groups. *Proc. Amer. Math. Soc. 57:* 25-28 (1976).

2. Y. Gurevich and W. C. Holland, Recognizing the real line. In press.

3. Michile Jambu-Giraudet, Theories des modeles de groupes d'automorphismes d'ensembles totalment ordonnes. *These 3me Cycle*, University of Paris VII (1979).

CLASSES WHICH GENERATE THE VARIETY OF ALL LATTICE-ORDERED GROUPS

George F. McNulty
University of South Carolina
Columbia, South Carolina

1. INTRODUCTION

A lattice-ordered group (ℓ-group) is *normal-valued* provided it satisfies the quasi-identity $\forall x,y[x \geq e \wedge y \geq e \rightarrow x^2y^2 \geq yx]$. (See Wolfenstein [10] for an alternative definition.) By a standard lattice theoretic argument, this quasi-identity can be replaced by an identity in the presence of the axioms for ℓ-groups and hence the class of normal valued ℓ-groups is a variety (alias equational class). In [7], W. Charles Holland proved that the variety of normal-valued ℓ-groups is the largest proper subvariety of the variety of all ℓ-groups. It follows that any class K of ℓ-groups which is provided with an ℓ-group which is not normal valued must generate the whole variety of ℓ-groups through the formation of homomorphic images of subalgebras of direct products of systems of members of K. By the use of a modest amount of model theory we refine this result of Holland's, chiefly by replacing the formation of direct product with the formation of ultraproducts. From his result, Holland deduced:

If $\mathcal{G}$ is a lattice-ordered group which is not normal valued and σ is an identity true in $\mathcal{G}$, then σ is true in every lattice-ordered group.

From the refinement presented here, the following stronger deduction can be made:

If $\mathcal{G}$ is a lattice-ordered group which is not normal valued and σ is a universal sentence true in $\mathcal{G}$, then σ is true in every lattice-ordered group.

2. SOME NOTATION AND A LITTLE MODEL THEORY

Our principal reference in model theory is Change and Keisler [3], while for notions from the general theory of algebraic structures we refer to Grätzer [4] and to Henkin, Monk, and Tarski [5, Chapter 0].

The *language* of lattice-ordered groups is distinguished by three binary operation symbols: $\wedge$ and $\vee$, denoting the lattice operations of meet and join, and $\cdot$, denoting the group composition, as well as one unary operation symbol: $^{-1}$, denoting the group inversion. The remaining symbols in the language are variables: x, y, z, w, u, v,..., the equality symbol: $\underset{\sim}{=}$, the symbol for "and": $\underset{\sim}{\wedge}$, the symbol for "and/or": $\underset{\sim}{\vee}$, the symbol for "not": $\underset{\sim}{\neg}$, and the symbol for "for every": $\underset{\sim}{\forall}$. Parentheses are used for punctuation. We also require certain expansions of this language. Let $\mathcal{G}$ be an ℓ-group. The *diagram language* of $\mathcal{G}$ has all of the symbols of the language of lattice-ordered groups and in addition it possesses a *constant symbol* $\underset{\sim}{g}$ for each $g \in G$, i.e. the diagram language of $\mathcal{G}$ has a name for each member of G. The (polynomial) terms, the formulas, and the sentences of our languages are constructed according to common mathematical usage (c.f. Chang and Keisler [3]). The *negative diagram* of the ℓ-group $\mathcal{G}$ is the set of all sentences of the form $\underset{\sim}{\neg}\phi \underset{\sim}{=} \psi$ which are true in $\mathcal{G}$ and in which ϕ and ψ are (necessarily) terms in which no variables occur. A sentence σ in the language of lattice-ordered groups is *positive universal* provided the negation symbol $\underset{\sim}{\neg}$ does not occur in σ, that is $\underset{\sim}{\forall}$, $\underset{\sim}{\vee}$, and $\underset{\sim}{\wedge}$ are the only logical quantifiers and connectives to occur in σ. ℓ-groups are denoted by capital script letters with the universe of an ℓ-group denoted by the corresponding capital Roman letter. An ℓ-group $\mathcal{G}$ is a *model* of a set of sentences provided all members of the set are true in $\mathcal{G}$. If $\mathcal{G}$ is an ℓ-group and Δ is a set of sentences from the diagram language of an ℓ-group $\mathcal{H}$, we say that $\mathcal{G}$ can be *expanded* to a model of Δ when for each $h \in H$, there is $g_h \in G$ such that all the sentences in Δ become true in $\mathcal{G}$ when $\underset{\sim}{h}$ is interpreted as g_h for each $h \in H$. If ϕ is a term in the language of ℓ-groups and $x_0,\ldots,x_{n-1}$ is a list of the variables occurring in ϕ, then in an ℓ-group $\mathcal{G}$, ϕ is interpreted as an n-place polynomial function, denoted by $\phi^{\mathcal{G}}$.

All the theorems from model theory that we require can be found in Chang and Keisler [3]. Other references are Grätzer [4] and Bell and Slomson [1]. We refer to the following theorems (references are from Chang and Keisler): the compactness theorem (1.3.22), Łos' theorem (4.1.9), the Keisler-Shelah ultrapower theorem (6.1.15), the Łos-Tarski theorem (3.2.2) and Lyndon's preservation theorem (3.2.4). Of these last two theorems we use only the easy parts.

We adhere to the following notation. Let K be a class of algebraic structures.

SK is the class of isomorphic images of substructures of members of K.

HK is the class of homomorphic images of members of K.

PK is the class of isomorphic images of direct products of systems of members of K.

$P_u K$ is the class of isomorphic images of ultraproducts of non-empty systems of members of K.

Let S be a set endowed with a linear order <. $\mathcal{A}(S)$ is the ℓ-group of all order preserving permutations of S. $\mathcal{A}(S)$ is called the *full ℓ-group* of S. $\mathcal{F}$ denotes the class of full ℓ-groups. Let $\mathcal{G}$ be an ℓ-subgroup of $\mathcal{A}(S)$. $\mathcal{G}$ is o-2-*transitive* provided that whenever $s_0, s_1, t_0, t_1 \in S$ with $s_0 < s_1$ and $t_0 < t_1$ there exists $g \in G$ such that $s_i g = t_i$ for $i = 0,1$. $\mathcal{T}$ denotes the class of o-2-transitive ℓ-groups. (It is known that every o-2-transitive ℓ-group is actually o-n-transitive, cf. [8].) $\mathcal{N}$ denotes the class of all normal-valued ℓ-groups and $\mathcal{V}$ denotes the class of all ℓ-groups.

3. GENERATING THE CLASS OF ALL ℓ-GROUPS

If K is any class of similar algebras, it is known that HSPK is the smallest equational class (i.e. variety) which includes K, see Birkhoff [2] and Tarski [9].

THEOREM. Let K be a class of ℓ-groups. The following are equivalent:

(1) HSPK = $\mathcal{V}$,

(2) Some member of K is not normal valued,

(3) HSK contains a non-trivial o-2-transitive ℓ-group,

(4) $HSP_u K = \mathcal{V}$.

Our proof requires four lemmas. The first lemma reflects the content of Lemma 3 in Holland's paper.

LEMMA 1. If $\mathcal{G}$ is a non-trivial o-2-transitive ℓ-group, $\mathcal{H}$ is any full ℓ-group and Δ is a finite subset of the negative diagram of $\mathcal{H}$, then $\mathcal{G}$ can be expanded to a model of Δ.

Proof. It is required to assign elements of G to the constant symbols which are names for elements of H and which occur in Δ, in such a manner that Δ becomes true in $\mathcal{G}$. Let $<S,\leq>$ be the linear ordering on which $\mathcal{G}$ is o-2-transitive. Since $\mathcal{G}$ is non-trivial, S must be infinite. Let $<S',\leq'>$ be the linear ordering on which $\mathcal{H}$ is the full ℓ-group. Now we can take

$$\Delta = \{\neg\phi_0 \approx \psi_0, \neg\phi_1 \approx \psi_1, \ldots, \neg\phi_{n-1} \approx \psi_{n-1}\}$$

where n is the number of sentences in Δ and $\phi_0,\phi_1,\ldots,\phi_{n-1}$ as well as $\psi_0,\psi_1,\ldots,\psi_{n-1}$ are terms (polynomials) in the diagram language of $\mathcal{H}$. Hence $\phi_i^{\mathcal{H}}$ and $\psi_i^{\mathcal{H}}$ are members of H and thus permutations on S', for all $i=0,1,\ldots,n-1$. Since $\neg\phi_i \approx \psi_i$ is true in $\mathcal{H}$ it must be that $\phi_i^{\mathcal{H}}$ and $\psi_i^{\mathcal{H}}$ are distinct permutations of S'. Consequently, for each $i=0,\ldots,n-1$ pick $t_i' \in S'$ so that $(t_i')\phi_i^{\mathcal{H}} \neq (t_i')\psi_i^{\mathcal{H}}$. Let

$$T' = \{(t_i')\theta^{\mathcal{H}}: \ i=0,1,\ldots,n-1 \text{ and } \theta \text{ is a subterm of } \phi_i \text{ or } \psi_i\}.$$

T' can be regarded as the set of all members of S' encountered in the computation of $(t_i')\phi_i^{\mathcal{H}}$ and $(t_i')\psi_i^{\mathcal{H}}$ for any i. T' is finite, since every term has only finitely many subterms. Since S is infinite it must have a subset T ordered by $\leq$ in the same way that T' is ordered by $\leq'$. (That is T' is order-embeddable in S.) Suppose T' (and hence T) has m members. Since $\mathcal{G}$ is o-2-transitive it is o-m-transitive. Suppose $h \in H$ and $\underset{\sim}{h}$ occurs in Δ; thus $\underset{\sim}{h}^{\mathcal{H}} = h$. Since $\mathcal{G}$ is o-m-transitive, G possesses a member g such that $g|T$ exhibits the same behavior as $h|T'$. So we assign g to $\underset{\sim}{h}$. Then if t_i corresponds to t_i', it must be so that

$$(t_i)\phi_i^{\mathcal{G}} \neq (t_i)\psi_i^{\mathcal{G}}$$

for all $i=0,1,\ldots,n-1$. [A completely detailed proof of this would demonstrate that $(t_i)\theta^{\mathcal{G}}$ corresponds to $(t_i')\theta^{\mathcal{H}}$ and the demonstration would proceed by induction on the complexity of the term θ.] Altogether we have described how to expand $\mathcal{G}$ to a model of Δ, which was as required.

LEMMA 2. If $\mathcal{G}$ is any non-trivial o-2-transitive ℓ-group and $\mathcal{H}$ is any full ℓ-group, then $\mathcal{H} \in \underset{\sim}{\mathbf{HSP}}_{\underset{\sim}{u}}\{\mathcal{G}\}$.

Proof. Let Σ be the set of all sentences in the diagram language of $\mathcal{G}$ which are actually true in $\mathcal{G}$. (Σ is called the *elementary diagram* of $\mathcal{G}$.) Let Γ be the negative diagram of $\mathcal{H}$. According to Lemma 1, every finite subset of $\Sigma \cup \Gamma$ has a model. By the compactness theorem for first-order logic $\Sigma \cup \Gamma$ itself has a model, say $\mathcal{C}$. $\mathcal{C}$ is an ℓ-group in which certain elements have been distinguished by symbolic names like $\underset{\sim}{g}$ and $\underset{\sim}{h}$ where $g \in G$ and $h \in H$. Since $\mathcal{C}$ is a model of Σ we conclude that $\mathcal{C}$ is an elementary extension of $\mathcal{G}$, (cf. theorem 3.1.3 in Chang and Keisler [3]). According to the Keisler-Shelah ultrapower theorem, $\mathcal{C}$ and $\mathcal{G}$ have isomorphic ultrapowers. This means that $\mathcal{G}$ has an ultrapower which can be expanded to a model of Γ. Let $\mathcal{D}$ be such an ultrapower of $\mathcal{G}$. Let $B = \{\phi^{\mathcal{D}}\colon\ \phi$ is a term in the diagram language of $\mathcal{H}$ such that no variables occur in $\phi\}$. It is easy to see that B is closed under the ℓ-group operations of $\mathcal{D}$ and so B specifies an ℓ-subgroup $\mathcal{B}$ of $\mathcal{D}$. Define $F(\phi^{\mathcal{D}}) = \phi^{\mathcal{H}}$ for all terms ϕ of the diagram language of $\mathcal{H}$ in which no variables occur. $F\colon B \to H$ since if $\phi^{\mathcal{D}} = \psi^{\mathcal{D}}$ then $\phi \underset{\sim}{=} \psi$ is true in $\mathcal{D}$ and so $\underset{\sim}{\neg}\phi \underset{\sim}{=} \psi$ is not in Γ; therefore $\phi^{\mathcal{H}} = \psi^{\mathcal{H}}$ since Γ is the negative diagram of $\mathcal{H}$. F is evidently a homomorphism from $\mathcal{B}$ onto $\mathcal{H}$. So $\mathcal{H} \in \mathbf{HSP}_{\mathbf{u}}\{\mathcal{G}\}$.

The last two lemmas are due to W. C. Holland.

LEMMA 3. (Holland [6]) $\mathcal{V} = \mathbf{S}\mathcal{F}$

LEMMA 4. (Holland [7, Lemma 2]) If $\mathcal{G}$ is an ℓ-group which is not normal-valued, then $\mathbf{HS}\{\mathcal{G}\}$ has a non-trivial o-2-transitive member.

PROOF OF THE THEOREM

We prove (1) => (2) => (3) => (4) => (1).

(1) => (2) Since the class of all normal-valued ℓ-groups is a proper subvariety of $\mathcal{V}$, if $\mathbf{HSP}K = \mathcal{V}$, then K cannot consist entirely of normal-valued ℓ-groups.

(2) => (3) is just Lemma 4.

(3) => (4) According to Lemma 2 $\mathcal{F} \subseteq \mathbf{HSP}_{\mathbf{u}}\mathbf{HS}K$. By Lemma 3 $\mathcal{V} \subseteq \mathbf{SHSP}_{\mathbf{u}}\mathbf{HS}K$. By some standard manipulations of $\mathbf{H}$, $\mathbf{S}$, and $\mathbf{P}_{\mathbf{u}}$ (in particular theorems 0.2.19, 0.3.12, and 0.3.69 in Henkin, Monk, and Tarski [5]) we obtain $\mathcal{V} \subseteq \mathbf{SHSP}_{\mathbf{u}}\mathbf{HS}K \subseteq \mathbf{HSP}_{\mathbf{u}}K$. Since $K \subseteq \mathcal{V}$ and $\mathcal{V}$ is closed under $\mathbf{S}$, $\mathbf{H}$, and $\mathbf{P}_{\mathbf{u}}$ (by virtue of being a variety), we conclude $\mathcal{V} = \mathbf{HSP}_{\mathbf{u}}K$.

(4) => (1) Since $\mathbf{P}_{\mathbf{u}}K \subseteq \mathbf{HP}K$ and $\mathcal{V}$ is a variety, the conclusion is immediate.

The theorem is established.

COROLLARY 1. No normal-valued non-trivial ℓ-group is o-2-transitive.

COROLLARY 2. If $\mathcal{G}$ is an ℓ-group which is not normal valued and σ is a positive universal sentence true in $\mathcal{G}$, then σ is true in every ℓ-group.

Proof. According to the theorem $\mathcal{V} = \underset{\sim}{H}\underset{\sim}{S}\underset{\sim}{P}_{\underset{\sim}{u}}\{\mathcal{G}\}$. By the theorem of Łos, σ is true in every member of $\underset{\sim}{P}_{\underset{\sim}{u}}\{\mathcal{G}\}$. According to the Łos-Tarski theorem and Lyndon's theorem, σ is true in all members of $\mathcal{V}$.

REFERENCES

1. J. L. Bell and A. B. Slomson, *Models and Ultraproducts*. North-Holland Publishing Company, Amsterdam, 1969.

2. G. Birkhoff, On the structure of abstract algebras. *Proc. Cambridge Philos. Soc.* 31: 483-454 (1935).

3. C. C. Chang and H. J. Keisler, *Model Theory*. North-Holland Publishing Company, Amsterdam, 1973.

4. G. Grätzer, *Universal Algebra*. Van Nostrand, Princeton, 1968.

5. L. Henkin, J. D. Monk, and A. Tarski, *Cylindric Algebras*, part I. North-Holland Publishing Company, Amsterdam, 1971.

6. C. Holland, The lattice-ordered group of automorphisms of an ordered set. *Michigan Math. J.* 10: 399-408 (1963).

7. C. Holland, The largest proper variety of lattice-ordered groups. *Proc. Amer. Math. Soc.* 57: 25-28 (1976).

8. S. H. McCleary, o-2-transitive ordered permutation groups. *Pacific J. Math.* 49: 425-429 (1973).

9. A. Tarski, Contributions to the theory of models, III. *Nederl. Akad. Wetensch. Proc. Ser. A.* 58: (Indag. Math. 17) 56-64 (1955).

10. S. Wolfenstein, Valeurs normales dans un groups réticulé. *Atti Accad. Naz. Lincei Rend Cl. Sci. Fis. Mat. Natur.* (8) 44: 337-342 (1968).

EQUATIONS AND INEQUATIONS IN LATTICE-ORDERED GROUPS

A. M. W. Glass
Bowling Green State University
Bowling Green, Ohio

Keith R. Pierce
University of Missouri-Columbia
Columbia, Missouri

Solving equations has long held a central position in the theory of mathematics. (People were even prepared to debauch in order to obtain the solution of the cubic!) It was realised fairly early on that it was often necessary to enlarge the field in order to obtain solutions. This great step forward -- and realising how the Greeks hated $\sqrt{2}$, it was a big step -- produced the idea of a field closed under solutions of polynomials. These were called algebraically closed fields. Such a field F had the added advantage that any finite set of polynomials which had a root in some extension field of F had a solution in F. This idea generalizes easily to groups, rings, ℓ-groups, etc.; viz: G is an *algebraically closed* group (ring, etc.) if any finite set of equations having a solution in a supergroup (ring etc.) over G has a solution in G. It is sometimes also helpful to simultaneously solve equations and inequations ($\neq$). In field theory, each inequation $p(X_1,\ldots,X_n) \neq 0$ can be replaced by an equation in one more unknown, namely $p(X_1,\ldots,X_n)Y = 1$. Hence algebraically closed fields further satisfy: any finite set of equations and inequations having a solution in an extension, has already a solution in the field. Unfortunately, this is not true in general (we will see that it is true for abelian o-groups, for o-groups, and for ℓ-groups but not for abelian ℓ-groups). This leads to the definition: A group (ring, ℓ-group, etc.) G is *existentially complete* if every finite set of equations and inequations having a solution in a supergroup (ring, etc.) has a solution in G. Obviously, algebraically closed and existentially complete objects are, in some sense, very full and desirable.

The purpose of this paper is to state some of the results that hold for ℓ-groups. We will rarely do more than sketch proofs, leaving a fuller write up till later.[1] We will often be imprecise, identifying classes of structures with their theories (e.g., groups $\leftrightarrow$ group theory), in the usual way. We have added an appendix on the model theory that we will need for our work. It should be consulted as the reader needs it. The results in the appendix are not new. We hoped that, by collecting them together here, some of the proverbial lazy algebraists who were interested would read them. The appendix is meant to be a brief, but hopefully comprehensive, account (with proofs) of the background material that motivated our work. It is self-contained and is intended to give the ordered-grouper a readily available, and we trust readable, account of some useful model theory.

The material in the first two sections was presented at the Boise meeting in two separate co-authored talks. The first "Existentially complete abelian ℓ-groups" was given by K.R. Pierce. The second "In search of the Holy Grail; existentially complete ℓ-groups" was given by A.M.W. Glass (who is grateful to the Bowing Green State University Faculty Committee for partially supporting his research). Both talks were dedicated to the memory of Abraham Robinson who is responsible for making possible the research on existentially complete structures.

We are extremely grateful to the following (in alphabetical order): Yuri Gurevich and W. Charles Holland for acting as penetrating listeners; William H. Wheeler for making us aware of the results about coding first and second order number theory into existentially complete division rings to distinguish between finitely and infinitely generic ones, (his suggestion that it might work for ℓ-groups, too, motivated part of section 2); Carol Wood for making us aware of the literature on existentially complete algebras (groups, nilpotent groups, division rings, etc.) without which we would never have tried to penetrate the partition of existentially complete classes into elementary equivalence classes.

The appendix is taken from the first author's lectures on model-theoretic forcing at the Michigan-Ohio Logic Seminar at the University of Michigan, Autumn 1978. All undefined terms in sections 1 and 2 appear in the appendix--we hope.

1. ABELIAN ℓ-GROUPS

LEMMA 1.1. The theory of divisible abelian o-groups is model-complete.

Proof Outline. This theory can be universally axiomatized if we adjoin to the language of ℓ-groups unary operations standing for rational multiplication. Therefore, by Lemma 0 of the appendix, it suffices to show: If $G \leq H$ are divisible abelian o-groups, $M \models \exists x\phi(x,g)$ where ϕ is quantifier-free and $g \in G$, then $G \models \exists x\phi(x,g)$. But ϕ is equivalent to a finite Boolean combination of atomic formulae of the form $x \leq a$ or $x = a$, $(a \in G)$. Thus x satisfies ϕ if and only if it lies in a union of finitely many intervals with endpoints from G. Since G is divisible, it is dense, and therefore this union is nonempty in H if and only if it is nonempty in G. Therefore $G \models \exists x\phi(x,g)$ as desired.

Since the class of abelian o-groups has the amalgamation property [15] then we have the

THEOREM 1.2. The theory of divisible abelian o-groups is the model completion of the theory of abelian o-groups; thus an abelian o-group is existentially complete (in the class of abelian o-groups) if and only if it is divisible.

Since divisibility is expressible by equations alone, then we have the

COROLLARY 1.3. The following are equivalent in the class of abelian o-groups:

(1) G is existentially complete

(2) G is algebraically closed

(3) G is divisible.

We turn now to the theory $\mathcal{A}$ of abelian ℓ-groups, where complications immediately arise:

THEOREM 1.4. Every direct sum of divisible abelian o-groups is $\mathcal{A}$-algebraically closed but not $\mathcal{A}$-existentially closed.

The proof of algebraic closedness uses Theorem 1.2 to construct solutions to systems of equations, one coordinate at a time. To show failure of existential completeness, consider the sentence $\exists x \exists y\ (x + y = g\ \&\ x \wedge y = 0\ \&\ x \neq 0\ \&\ y \neq 0)$. This is false for any basic element of an abelian ℓ-group but is true in an extension; just embed the ℓ-group in the direct sum of two copies of itself via the diagonal map.

By using these sorts of arguments one can easily prove the

PROPOSITION 1.5. If G is $\mathcal{A}$-existentially complete, then

(1) G has no basic or special elements

(2) G is divisible

(3) G has no weak order unit.

However the most important property, as we shall see, is described next. We say that $a \in G$ *splits over* $b \in G$ if there are disjoint strictly positive elements $x,y \in G$ such that $a = x + y$ and $y \wedge b = 0$,

LEMMA 1.6. If $0 < a, b \in G$ then the following are equivalent:

(1) a splits over b in some abelian extension H

(2) $a \notin G(b)$.

Proof. (1) $\to$ (2) is obvious. Conversely, if $a \notin G(b)$ then there is a prime subgroup P of G such that $b \in P$ and $a \notin P$. If $H = G \boxplus G/P$, then $x = (a,0)$ and $y = (0,a+P)$ satisfy the splitting property.

PROPOSITION 1.7. If G is $\mathcal{A}$-existentially complete, then for every $0 < a, b \in G$, a splits over b if and only if $a \notin G(b)$.

THEOREM 1.8. If G is an existentially complete abelian ℓ-group, the following properties are equivalent to first order properties:

(1) $a \in G(b)$

(2) $a \ll b$

(3) G is archimedean

(4) G is hyperarchimedean.

Proof. (1) is given by Proposition 1.7. It follows then that $a \ll b$ if and only if $\forall x\ (x \in G(a) \to x < b)$. (3) now follows immediately. (4) is also immediate once we recall that G is hyperarchimedean if and only if $G = G(g) \boxplus g^{\perp}$ for every $0 < g \in G$.

COROLLARY 1.9. $\mathcal{A}$ has no model companion. Hence the class of existentially complete abelian ℓ-groups is not an elementary class.

Proof. By the remarks in the appendix, it suffices to find an $\mathcal{A}$-existentially complete group, some ultrapower of which is not $\mathcal{A}$-existentially complete. Let G be $\mathcal{A}$-existentially complete and fix $0 < g \in G$. Let I be the set of positive integers, $\mathcal{U}$ be a non-principal ultrafilter on I, and let $H = \Pi G/\mathcal{U}$. Finally, let $h_1 = (g,g,g,\ldots)^{\sim}$, $h_2 = (g,2g,3g,\ldots)^{\sim}$. At each coordinate the first order property "h_2 splits over h_1" is true, so the same is true in H. But in H, $h_2 \notin H(h_1)$, so by Proposition 1.7, H cannot be $\mathcal{A}$-existentially complete.

Now abelian ℓ-groups are precisely the subdirect sums of abelian o-groups. Thus Theorem 1.2 and Corollary 1.9 kill the conjecture of W. H. Wheeler's [17]: If T is a companionable universal theory then so is the theory of the class of subdirect sums of models of T.

We next delve a bit deeper into the complexity of the class of existentially complete abelian ℓ-groups. Referring to the appendix, Lemmas 10 and 18, we know that the classes of finitely generic and infinitely generic abelian ℓ-groups are classes of $\mathcal{A}$-existentially complete ℓ-groups. We show that these classes are distinguishable by an $\forall\exists\forall$ sentence, the easiest possible by the appendix, Lemma 22; that is, there is a first order sentence $\phi = \forall \tilde{x}\, \exists \tilde{y}\, \forall \tilde{z}\, \psi(\tilde{x},\tilde{y},\tilde{z})$, where ψ contains no quantifiers, with the property that ϕ is true in all finitely generic abelian ℓ-groups but no infinitely generic ones. The sentence is the one which expresses hyperarchimedeanness.

THEOREM 1.10. No infinitely generic abelian ℓ-group is archimedean.

Proof. By Theorem 7 of the appendix there are certainly non-archimedean infinitely generic abelian ℓ-groups. But they are all elementarily equivalent (Lemma 11 of the appendix). Therefore none are archimedean.

THEOREM 1.11. Every finitely generic abelian ℓ-group is hyperarchimedean.[2]

The proof needs the definition and properties of $\overline{C}(X,\mathbb{R})$, which we now describe: Let X be a topological space, G an abelian o-group endowed with the discrete topology, and C(X,G) the ℓ-group of continuous functions from X to G having compact support.

LEMMA 1.12. If X is locally compact, totally disconnected, perfect, and Hausdorff, but not compact, then $\overline{C}(X,\mathbb{R})$ is an $\mathcal{A}$-existentially complete hyperarchimedean ℓ-group.

The proof will be published elsewhere.[3] For future reference, we note that, under these hypotheses, the elements of $\overline{C}(X,\mathbb{R})$ are precisely those functions f having finite range and such that for every nonzero r in the range, $f^{-1}(r)$ is a compact open subset of X. We do not know if this example is finitely generic; we conjecture that it is.

In view of this example, one might expect that if G is $\mathcal{A}$-existentially complete and hyperarchimedean, then it is representable as a group of real-valued functions with finite range. This is false, as the next example shows:

EXAMPLE 1.13. Let X be the disjoint union of denumerably many Cantor sets X_n, $1 \leq n < \omega$, so that X satisfies the conditions of Lemma 1.12. Let $f:X \to \mathbb{R}$ be defined by setting $f(x) = \pi + 1/n$ whenever $x \in X_n$. Let H be the ℓ-group generated by C(X,Q) together with f. Finally, let G be the direct sum of denumerably many copies of H. Then G is $\mathcal{A}$-existentially complete, but is not representable as an ℓ-group of real-valued functions with finite range, as is shown in Example 7.1 of [4].

Our final result concerning abelian ℓ-groups is

THEOREM 1.14. There are continuum many mutually non-isomorphic countable existentially complete abelian ℓ-groups.

Proof. It is an easy matter to construct continuum many non-isomorphic rank 2 divisible subgroups D of $\mathbb{R}$. Each of these is embeddable in a countable existentially complete abelian ℓ-group, and the theorem now follows immediately. Interestingly, however, we can explicitly display continuum many such; Let X be the Cantor set, and let B be any countable Boolean algebra of clopen subsets of X. For each D above, let G(D) be the ℓ-subgroup of $\overline{C}(X,D)$ consisting of those f for which $f^{-1}(d) \in B$ for all $d \in D$. Then G(D) is $\mathcal{A}$-existentially complete, countable, and $G(D) \cong G(E)$ if and only if $D = E$.

We now turn to some examples of companionable theories of abelian o-groups and ℓ-groups. Here we will be content with statements of results and hints at their proof. First of all, we enlarge the language of ℓ-groups by adding a new constant symbol 1 which will be named a "unit" in these groups. The main effect this has is to restrict ℓ-homomorphisms to those that preserve this unit.

Recall that an o-group is *discrete* if it has a smallest positive element -- equivalently, if it has a smallest proper convex subgroup isomorphic to $\mathbb{Z}$; indeed, we shall identify this convex subgroup with $\mathbb{Z}$. In particular, 1 always names the smallest positive element. A discrete o-group G is *regularly discrete* if $|G/pG| = p$ for every prime number p. These groups were studied by Robinson and Zakon [16], and they showed that G is regularly discrete if and only if, for every prime number p, every element of G is congruent, modulo p, to one of $0,1,2,\ldots,p-1$. Conrad [3] later showed regularity equivalent to $G/\mathbb{Z}$ being divisible. Let $\mathcal{D}$ and $\mathcal{D}^*$ denote (the first order theories of) the respective classes of discrete and regularly discrete abelian o-groups.

THEOREM 1.15. $\mathcal{D}^*$ is the model companion of $\mathcal{D}$.

THEOREM 1.16. $\mathcal{D}$ does not have the amalgamation property, so $\mathcal{D}$ has no model completion.

Call an ℓ-group *discrete* if it is a subdirect product of discrete o-groups. Since we are naming 1, this forces any discrete ℓ-group to contain the function which is constantly 1. By general model theory, the class of discrete ℓ-groups is elementary, being axiomatized by the set of universal Horn sentences which are deducible from the theory $\mathcal{D}$. However, in this case, an axiomatization can be explicitly displayed, namely by listing the axioms for the theory of abelian ℓ-groups, together with the statements that 1 is positive, singular, and a weak order unit. Certainly any discrete abelian ℓ-group has these properties; conversely, if G satisfies these axioms, then the minimal prime subgroups of G are exactly the prime subgroups missing 1. For such primes P, G/P is discrete and therefore $G \leq \Pi\{G/P \mid P \text{ minimal prime}\}$ represents G as a subdirect sum of discrete o-groups. Denote by $\mathcal{DL}$ the theory of discrete abelian ℓ-groups.

Call a discrete ℓ-group *regularly discrete* if G/S is divisible, where S is the ℓ-ideal generated by the singular elements of G. As in the totally ordered case, the class of regularly discrete ℓ-groups can be axiomatized by adjoining to $\mathcal{DL}$ the statements, one for every prime number p, that every element of G is congruent modulo p to a sum of at most p(p - 1)/2 singular elements. That this characterize the theory $\mathcal{RDL}$ of regularly discrete ℓ-groups is readily verified once one observes that a regularly discrete ℓ-group is a subdirect sum of regularly discrete o-groups.

Finally, let $\mathcal{DL}^*$ denote the theory obtained by adjoining to $\mathcal{RDL}$ the following statements: (1) there are no minimal singular elements, and (2) for every $0 < b$ and every singular s, there exist x and y such that $b = x + y$, $x \wedge y = 0$, $y \wedge s = 0$, and $x \wedge (1 - s) = 0$. Viewing G as a subdirect sum of discrete o-groups, this axiom states that b splits into two summands one of which equals b on the support of s and the other on the support of 1 - s.

It is apparent that any $\mathcal{DL}$-existentially closed ℓ-group must be a model of $\mathcal{DL}^*$. In fact, this is sufficient:

THEOREM 1.17. $\mathcal{DL}^*$ is the model companion, but not the model completion, of $\mathcal{DL}$.

$\mathcal{DL}$ fails to have the amalgamation property for the same reason that it fails in $\mathcal{D}$.

A word about the proof of this theorem:

Lipschitz and Saracino [11] have determined the model companion of the theory of semiprime commutative rings -- it is, speaking loosely, the theory of algebraically closed von Neumann regular rings without minimal idempotents. The proof of our theorem 1.17 closely parallels their proof; in fact, one can set up a dictionary for translating ring theory terms into discrete ℓ-group terms (for example, idempotent = singular element, integral domain = discrete o-group, algebraically closed field = regularly discrete o-group, etc.), and after translation, their proof looks like ours (warning: there are some complications if one really attempts this!). For a very transparent proof of Lipschitz and Saracino's result, see Cherlin [2, pp 96-107].

In order not to leave the reader totally dangling, we sketch an outline of the proof. Suppose G, $H \models \mathcal{DL}^*$, $G \leq H$, $H \models \exists \tilde{x}\, \phi(\tilde{x},\tilde{a})$. The axioms for $\mathcal{DL}$ imply that G is an ℓ-group of functions defined on a Stone space X (the space associated with the Boolean algebra of singular elements of G) and taking values in various regularly discrete o-groups. Using Theorem 1.15 one finds a finite partition $\{U_i\}$ of X and associated elements $\{g_U\}$ of G such that every atomic component of ϕ is true for all g_U, and for every negated atomic component of ϕ there is some i such that that component is true for g_{U_i} at every point of U_i. Axiom (2) of $\mathcal{DL}^*$ allows us to patch together these g_U, creating a solution of ϕ in G.

LEMMA 1.18. The theories $\mathcal{RD}$ and $\mathcal{RDL}$ have the amalgamation property.

THEOREM 1.19. $\mathcal{DL}^*$ is the model completion of $\mathcal{RDL}$.

COROLLARY 1.20. $\mathcal{DL}^*$ is a complete and decidable theory.

Proof. Since $\mathbb{Z} \in \mathcal{RDL}$ and is canonically contained in every member of $\mathcal{DL}^*$, the amalgamation property for $\mathcal{RDL}$ together with the mutual model consistency of $\mathcal{RDL}$ and $\mathcal{DL}^*$ imply $\mathcal{DL}^*$ enjoys the joint embedding property (any two models are simultaneously embeddable in a third). It is immediate that completeness follows from model-completeness and the joint embedding property. Since $\mathcal{DL}^*$ is recursively axiomatizable and complete, it is decidable.

2. EXISTENTIALLY COMPLETE LATTICE-ORDERED GROUPS

LEMMA 2.1. Every existentially complete ℓ-group G is divisible, and any two positive elements of G are conjugate. Hence existentially complete ℓ-groups are ℓ-simple.

This follows since the existence of nth roots can be said existentially and any ℓ-group can be ℓ-embedded in a divisible one. Also g_0 is conjugate to g_1 can be said existentially. Moreover, every ℓ-group can be ℓ-embedded in one in which any two positive elements are conjugate. Actually, Lemma 2.1 can be strengthened to: every existentially complete ℓ-group is simple as a group.

Either using the ℓ-embeddings mentioned above, or by Theorem 3 in the appendix:

THEOREM 2.2. Every non-trivial algebraically closed ℓ-group is existentially complete.

Existentially complete ℓ-groups are very complicated. Each generates $\mathcal{L}$ as a variety (the statement "G is not normal-valued" is existential). They have no basic elements, no special elements (if there were one, all positive elements would be special, by Lemma 2.1, which would imply the ℓ-group was an o-group), and no finite maximal pairwise disjoint set of elements. They are not finitely generated nor finitely related. Also, every element of an existentially complete ℓ-group is a commutator. Macintyre has shown further (see Corollary 24 of the appendix) that a finitely generated ℓ-group can be ℓ-embedded in every existentially complete ℓ-group only if it has soluble word problem. So telling apart different existentially complete ℓ-groups by looking at their ℓ-subgroups will be difficult (we would have to use ℓ-groups with insoluble word problem); moreover, there are an abundance of different existentially complete ℓ-groups:

THEOREM 2.3. There are $2^{\aleph_0}$ pairwise non-isomorphic countable existentially complete ℓ-groups.

For the proof, see Theorem 1.14.

Since existentially complete ℓ-groups are ℓ-simple, they have a transitive realization in some $A(\Omega)$, the ℓ-group of all order-preserving permutations of a totally ordered set Ω, with $f \leq g$ if $\alpha f \leq \alpha g$ for all $\alpha \in \Omega$. We might wonder what kind of realization can be achieved. We don't know the complete answer to this. All we know is

(1) if an existentially complete ℓ-group has an o-primitive representation, this representation must be pathologically o-2 transitive.

(2) if (1) fails for an existentially complete ℓ-group, G has a locally o-primitive representation and every element of G fixes infinitely many $\mathscr{C}$ classes (not necessarily pointwise) and moves an unbounded collection of $\mathscr{C}$ classes for any convex G-congruence $\mathscr{C}$ with $\mathscr{C} \neq \Omega \times \Omega$.

In either case (and we suspect only the latter) we have new (previously unknown) (ℓ-)simple ℓ-groups. If any satisfy (2), they will be the first known ℓ-simple ℓ-groups not to have an o-primitive representation and yet not be o-groups. (Very little is known about ℓ-simple ℓ-groups; we know of no example, even, of a normal valued ℓ-simple ℓ-group that is not an o-group).

In a fashion similar to the abelian case, to show that the theory of ℓ-groups has no model companion, we seek out a property that is not usually first order but which when restricted to existentially complete ℓ-groups becomes first order. As opposed to the abelian case, the theory is enriched by the ability to consider commutators. $g \in \langle h \rangle$ stands for "g belongs to the subgroup generated by h"; i.e., the sentence $\bigvee_{n=-\infty}^{\infty} (g = h^n)$ of $\mathscr{L}_{\omega_1\omega}$ ($\exists n$ is, of course, illegal). Let $C(g) = \{x \in G: xg = gx\}$, the centraliser of g. $C(h) \subseteq C(g)$ can be said first order: $\forall x(xh = hx \to xg = gx)$. Fortuitously:

LEMMA 2.4. If G is an existentially complete ℓ-group and $e < g, h \in G$, then $g \in \langle h \rangle$ if and only if $G \models C(h) \subseteq C(g)$ (so $g \in \langle h \rangle$ can be said by an $\forall$ sentence for existentially complete ℓ-groups).

This is the same as in group theory but the proof is very different. Left to right is obvious. Right to left in groups proceeds painlessly via the free product with amalgamation. No such tool is available to us for ℓ-groups and we have to resort to a messy computational proof via ordered permutation groups. First note that if $G \subseteq \mathcal{A}(\Omega)$ and there exists $x \in \mathcal{A}(\Omega)$ such that $x \in C(h) \setminus C(g)$, then the same is true in G since this is an existential statement. The lemma is proven by first establishing that $C(h) \subseteq C(g)$ implies that every interval of support of g is a supporting interval of h, and that on any supporting interval Δ of h, there is $n \in \mathbb{Z}$ such that $g|\Delta = h^n|\Delta$. The last part, to establish that n is independent of Δ, is rather technical and uses the fact that we can take the set of supporting intervals of g to be an α-set of degree 2 for suitable α.

The consequences of Lemma 2.4 are many. In fact, all the remaining results of this section hinge upon it.

THEOREM 2.5. The theory of ℓ-groups has no model-companion.

The proof is cheap, and mirrors that of Corollary 1.9. If there were a model-companion, $\Pi G/\mathcal{U}$ would be existentially complete for every existentially complete ℓ-group G and every ultrafilter $\mathcal{U}$. But if $h = (g,g,g,\ldots)^{\sim}$ and $f = (g,g^2,g^3,\ldots)^{\sim}$ then $C(h) \subseteq C(f)$ (by Łos' Theorem since it's true at every coordinate), and $f \notin \langle h \rangle$ if $\mathcal{U}$ is non-principal. This contradicts 2.4.

Using Lemma 2.4 again, we can now code first order arithmetic into our theory as follows. Let G be existentially complete and let $e < g \in G$.

$$g^n \oplus g^m = g^p \leftrightarrow g^n g^m = g^p \text{ (i.e., } n + m = p) \text{ and}$$
$$g^n \otimes g^m = g^p \leftrightarrow (\exists x)(x^{-1}gx = g^m \;\&\; x^{-1}g^n x = g^p) \quad \text{(i.e., } nm = p)$$

Since any two positive elements of G are conjugate, the model of arithmetic so obtained is independent of the choice of $e < g \in G$. Hence, courtesy of K. Gödel and Lemma 20 of the appendix:

THEOREM 2.6. The theory of finitely generic ℓ-groups is undecidable. Moreover, the theory is hyperarithmetical (i.e., as complicated as first order arithmetic).

COROLLARY 2.7. The group theory part of the theory of ℓ-groups is undecidable.

This compares nicely with Y. Gurevich's result that the lattice theory of the theory of abelian (and hence all) ℓ-groups is undecidable [9].

THEOREM 2.8. In any existentially complete ℓ-group, for any positive g, there is an h such that$\ldots \ll h^2gh^{-2} \ll hgh^{-1} \ll g \ll h^{-1}gh \ll h^{-2}gh^2 \ll \ldots$ where $x \ll y$ means $x^n \leq y$ for all positive integers n.

This shows how far from being archimedean these ℓ-groups are. (Since they are non-abelian, they are non-archimedean, but we had no knowledge how badly they would fail to be non-archimedean. Of course, we'd expect something bad, so Theorem 2.8 should be no surprise.)

Let $S(x_0,x_1)$ be the formula: $(x_1^{-1}x_0x_1 \wedge x_0 = e)$ & $(x_0 > e)$ & $(x_1 > e)$.

LEMMA 2.9. Let $X \subseteq \mathbb{Z}^+$. Then there exist $g_0, g_1 \in \mathcal{A}(\mathbb{R})$ such that $\mathcal{A}(\mathbb{R}) \models S(g_0, g_1^n)$ if and only if $n \in X$.

There is an infinitely generic ℓ-group G containing $\mathcal{A}(\mathbb{R})$ as an ℓ-subgroup (Theorem 7 of the appendix). $G \models S(g_0, g_1^n)$ if and only if $n \in X$ for some $g_0, g_1 \in G$ (g_0, g_1 dependent on X). Since the theory of ℓ-groups enjoys the joint embedding property, if H is any infinitely generic ℓ-group and X is any subset of $\mathbb{Z}^+$, there are $h_0, h_1 \in H$ such that $H \models S(h_0, h_1^n)$ if and only if $n \in X$ (Lemma 11 of the appendix). Using this and Lemma 2.4, we can code second order arithmetic into the theory of infinitely generic ℓ-groups. So:

THEOREM 2.10. The theory of infinitely generic ℓ-groups is Δ_1^2 (i.e., as complicated as second order number theory).

COROLLARY 2.11. The theories of infinitely generic ℓ-groups and finitely generic ℓ-groups share no models in common. Indeed, the theories can be distinguished by an $\forall\exists\forall$ sentence.

The first part follows from Theorem 2.10 and the fact that both theories are complete (Appendix Lemmas 11 & 21). The second part follows by general considerations ([10] chapter 16; for other parallels, see Macintyre's excellent survey article [14]).

COROLLARY 2.12. There are $2^{\aleph_0}$ pairwise non-elementarily equivalent countable existentially complete ℓ-groups. (cf Theorem 2.3)

Again, this follows from general considerations (see [10] chapter 7).

The $\forall\exists\forall$ sentence guaranteed by Corollary 2.11 is unsatisfactory in one sense: It involves coding Gödel number of sentences, proofs, etc., and thus is not meaningful to an algebraist. It would be very satisfying to find any algebraic sentence, such as the $\forall\exists\forall$ sentence that reflects the archimedean property in the abelian case. One possibility might be the transitive representation discussed after Theorem 2.3. However, we know of no way to distinguish the representations (1) and (2) above for existentially complete ℓ-groups. It may be impossible for all we know.

Considering the relation $x_0 \ll x_1$ again we note that, as in the abelian case, it is equivalent to a first order formula, namely $\forall x_2(x_2 \in \langle x_0\rangle \rightarrow x_2 \leq x_1)$, where $x_2 \in \langle x_0\rangle$ is the $\forall$ formula given by Lemma 2.4. This $(\exists x_0 > e)(\exists x_1 > e)(x_0 \ll x_1)$ is an $\exists\forall\exists$ sentence. Since existentially complete ℓ-groups are non-abelian, and archimedean ℓ-groups are abelian, this sentence holds in all existentially complete ℓ-groups. But it is easy to see (without recourse to big theorems) that existentially complete ℓ-groups are non-archimedean since $x_0 \wedge x_1^{-1}x_0x_1 = e$ & $x_1 > e$ imply $x_0 \ll x_1$ in any ℓ-group.(*) $\exists x_0 \exists x_1$ $(x_0 \wedge x_1^{-1}x_0x_1 = e$ & $x_1 > e)$ holds in $\mathbb{Z}\mathrm{wr}\mathbb{Z}$ and hence in $H \boxplus (\mathbb{Z}\mathrm{wr}\mathbb{Z})$ for any ℓ-group H. So if H is existentially complete, the $\exists$ sentence (*) holds in H.

What about the relation "$x_1^{-n}x_0x_1^n \wedge x_0 = e$ for all $n \in \mathbb{Z}^+$, $x_1 > e$ and $x_0 > e$"? This can be said first order in existentially complete ℓ-groups, viz: $\forall x_3(x_3^{-1}x_0x_3 \wedge x_0 = e$ or $x_3 \notin \langle x_1\rangle$ or $x_3 \leq e)$; i.e., $\forall x_3 \exists x_4$ $(x_3^{-1}x_0x_3 \wedge x_0 = e$ or $x_3 \leq e$ or $(x_4x_1 = x_1x_4$ & $x_4x_3 \neq x_3x_4))$. Hence $\exists x_0 \exists x_1$ $(x_0 > e$ & $x_1 > e$ & $x_1^{-n}x_0x_1^n \wedge x_0 = e$ for all $n \in \mathbb{Z}^+)$ can be said $\exists\forall\exists$, so is sufficiently complicated that it might hold in only some existentially complete ℓ-groups (it holds in infinitely generic ones). Also, we wouldn't expect to be able to ensure x_0 is disjoint from its conjugates by all positive powers of x_1 in a finite amount of information (which is all that is available for finitely generic structures) for all positive x_1. But the information can be captured from the following: $x_0 > e$ & $x_1 > e$ & $x_1^{-1}x_0x_1 \leq x_0$ & $x_1^{-1}x_2x_1 \leq x_2$ & $x_2 \wedge x_0 = e$, since it implies $x_1^{-n}x_0x_1^n \leq x_2$ for all $n \in \mathbb{Z}^+$ (and hence $x_1^{-n^2}x_0x_1^n \wedge x_0 = e$ for all $n \in \mathbb{Z}^+$). So the existence of such x_0, x_1 is guaranteed in every existentially complete ℓ-group. However, it is clear (by a routine (logic) compactness argument) that the ℓ-group with presentation $(x_0, x_1\colon\ x_1 \wedge e = e,\ x_1^{-1}x_0x_1 \wedge x_0 = e,\ x_1^{-2}x_0x_1^2 \wedge x_0 = e, \ldots)$ can not be finitely presented. Hence:

THEOREM 2.13. There is a finitely presented ℓ-groups which has an ℓ-group that is recursively presented but can not be finitely presented.

Recall Higman's Theorem: Every recursively presented (finite number of generators, recursively enumerable set of relations) can be embedded in a finitely presented group. Were this true for ℓ-groups, we would have to try to duplicate Macintyre's work for groups (see [13] and [14]) without the amalgamation property which would be rather difficult. But the lack of knowledge of an analogue for Higman's theorem clearly points out the trouble we have in coming up with an "algebraic" sentence to distinguish between finitely and infinitely generic ℓ-groups. If the analogue of Higman's theorem holds, then using [6] or [7] (or see [8]), we'd have a finitely presented ℓ-group with insoluble word problem. If the analogue of Higman's theorem failed, then we would be in a good position to get at a sentence which held in infinitely generic but not finitely generic ℓ-groups. (The crux being that a forcing condition is a finite set of equations and inequations. The inequations would be replaced by further equations giving rise to a finite set of equations in a finite number of generators; i.e., a finite presentation. Hence no condition would force the hoped-for sentence of the recursively presented ℓ-group which can't be ℓ-embedded in any finitely presented ℓ-group). However, one possible candidate is that there are $x_0, x_1 > e$ such that for any $n \in \mathbb{Z}^+$, there exist $m_1, m_2 > n$ such that $x_1^{-m_1}x_0x_1^{m_1} \wedge x_0 = e$ & $x_1^{-m_2}x_0x_1^{m_2} \wedge x_0 \neq e$, and yet this doesn't happen for m_1, m_2 belonging to some cyclic subgroups of $\mathbb{Z}$ (i.e., no easy escape like $x_1^{-n}x_0x_1^n = x_0$ if $n \in \mathbb{Z}^+$ is even and $x_1^{-n}x_0x_1^n \wedge x_0 = e$ if $n \in \mathbb{Z}^+$ is odd, which can be deduced from $x_1^{-2}x_0x_1^2 = x_0$ & $x_1^{-1}x_0x_1 \wedge x_0 = e$). The sentence which says this for existentially complete ℓ-groups is:

$$\exists x_0 \exists x_1 \forall x_2 \exists x_3 \exists x_4 \exists x_5 \forall x_6 \forall x_7 \forall x_8 \forall x_9 \exists x_{10} \exists x_{11} \forall x_{12} \forall x_{13} \exists x_{14} \exists x_{15} \exists x_{16} \exists x_{17}\ \phi(x_0, \ldots, x_{17})$$

where $\phi(x_0,\ldots,x_{17})$ is:

$$(x_0 > e) \& (x_1 > e) \& \{(x_2x_3 \neq x_3x_2 \& x_1x_3 = x_3x_1) \text{ or } [(x_4 > x_2 \& x_5 > x_2 \& x_4^{-1}x_0x_4 \wedge x_0 = e \& x_5^{-1}x_0x_5 \wedge x_0 \neq e) \& (x_6x_4 = x_4x_6 \text{ or } x_6x_1 \neq x_1x_6) \& (x_7x_5 = x_5x_7 \text{ or } x_7x_1 \neq x_1x_7)]\} \& \{[(x_{10} > x_8 \& x_{10}^{-1}x_0x_{10} \wedge x_0 = e) \& (x_{12}x_{10} = x_{10}x_{12} \text{ or } x_{12}x_1 \neq x_1x_{12}) \& (x_{14}x_8 = x_8x_{14} \& x_{10}x_{14} \neq x_{14}x_{10})] \text{ or } (x_{16}x_1 = x_1x_{16} \& x_{16}x_8 \neq x_8x_{16})\} \& \{[(x_{11} > x_9 \& x_{11}^{-1}x_0x_{11} \wedge x_0 \neq e) \& (x_{13}x_{11} = x_{11}x_{13} \text{ or } x_{13}x_1 \neq x_1x_{13}) \& (x_{15}x_9 = x_9x_{15} \& x_{15}x_{11} \neq x_{11}x_{15})] \text{ or } (x_{17}x_1 = x_1x_{17} \& x_{17}x_9 \neq x_9x_{17})\}$$

4

This $\exists\forall\exists\forall\exists\forall\exists$ sentence is algebraic but we have no idea if it fails to hold in finitely generic ℓ-groups. (It holds in infinitely generic ℓ-groups.) The trouble is to be sure that certain finite sets of basic sentences of $\mathcal{L}(C)$ are consistent with the theory of ℓ-groups. As the example leading up to Theorem 2.13 illustrates, considerable caution has to be exercised. (The condition $p = \{c_0 > e,\ c_1 > e,\ c_2 \wedge c_0 = e,\ c_1^{-1}c_2c_1 \leq c_2,\ c_1^{-1}c_0c_1 \leq c_2\}$ cannot be augmented to $\{c_0 > e,\ c_1 > e,\ c_2 \wedge c_0 = e,\ c_1^{-1}c_2c_1 \leq c_2,\ c_1^{-1}c_0c_1 < c_2,\ c_1^{-n}c_0c_1^{n} \wedge c_0 \neq e\}$ for any $n \in \mathbf{Z}^+$ and obtain a consistent set of basic sentences. If one tries $\Vdash \neg\exists\forall\exists$ sentence, taking p as indicated above, a contradiction arises since p cannot be extended as we would like.)

There are two other possibilities to try for "algebraic" sentences. The first is "complete distributivity." We know by Theorem 2.1 that $D(G) = G$ or $\{e\}$. Maybe it equals G in some existentially complete ℓ-groups and $\{e\}$ in others. (The representation (1) discussed after Theorem 2.3 would yield the former.) The trouble is in describing complete distributivity first order for existentially complete ℓ-groups. The second is "laterally complete." R. N. Ball has shown that if in some ℓ-group G, $\{g_i | i \in I\}$ are pairwise disjoint and $g = \bigvee_{i\in I} g_i$ exists in G, then there is an ℓ-group H containing G as an ℓ-subgroup such that $\bigvee_{i\in I} g_i$ exists in H but doesn't equal g (see [1] for the method). Using this fact and a certain definable set of $\{g_i | i \in I\}$ whose supremum, if it existed, in an existentially complete ℓ-group would be first order definable, we can obtain (and the reader is not expected to be able to fill in the gaps since none of this should appear to be at all first order (infinite suprema aren't in the language)):

THEOREM 2.14. No existentially complete ℓ-group is laterally complete.

Consequently, we are still unable to come up with an algebraic sentence distinguishing finitely generic ℓ-groups from infinitely generic ones. Moreover, getting at D(G), even if we can do so, may not help since first order definable suprema and infima may exist so infrequently that they give no clue at all about complete distributivity.

APPENDIX:

THE LAZY ALGEBRAIST'S GUIDE TO MODEL-THEORETIC FORCING

The purpose of these notes is to provide a mainly self-contained exposition of the model theory needed to understand the results of this paper. In order to keep the length down so that the proverbial lazy algebraist would read these notes, we have occasionally been miserly with our explanations. A very complete and thorough account can be found in [10]. Unless otherwise stated, the concepts and general results on model companions, existentially complete structures, and finitely and infinitely generic structures are due to Abraham Robinson and his collaborators.

Throughout, $\mathcal{L}$ will be a countable language including the equality relation (e.g., the language of ordered fields would have $+,-,\cdot$ as binary operations, $^{-1}$ as a unary operation, $\leq$ as a binary relation, and constants 0 and 1) and a countable list of variables $x_0, x_1, x_2, \ldots$. We build up terms from operations, constants and variables as usual (e.g., $1 + (0 \cdot 1)$, $(1 + 1) \cdot (1 - x_0))$. Atomic formulae are equations between terms, or relations applied to terms. Formulae are built up from atomic formulae using $\neg$ (not), & (and), or and $\exists$ (there exists). $\forall x\phi(x)$ will be short for $\neg \exists x \neg \phi(x)$. A sentence is a formula with no free variables.

A *theory* in $\mathcal{L}$ is a consistent set of sentences.

A *structure* in $\mathcal{L}$ is a set A together with interpretations in A for the relations, operations, and constants of $\mathcal{L}$; for our canonical example, a structure $\mathcal{A}$ would be a set A together with interpretations for $+,-,\cdot,^{-1},\leq,0$, and 1 in A. We will use $\mathcal{A},\mathcal{B},\ldots$ for structures.[5] Further, the underlying set for $\mathcal{A}$ will be denoted by A.

If $\phi(\underset{\sim}{x})$ is a formula in $\mathcal{L}$, we will define $\mathcal{A}$ satisfies $\phi(\underset{\sim}{x})$ at $\underset{\sim}{a} \in \mathcal{A}$ $(\mathcal{A} \models \phi(\underset{\sim}{a}))$ inductively.

1. If $\phi(\underset{\sim}{x})$ is atomic, $\mathcal{A} \models \phi(\underset{\sim}{a})$ if and only if $\phi(\underset{\sim}{a})$ is true in $\mathcal{A}$ (e.g., if $\phi(\underset{\sim}{x})$ is $(1 + 1)\cdot(1 - x_0) \leq 1 + x_0$, then $\mathcal{A} \models \phi(\underset{\sim}{a})$ if and only if $(1 + 1)(1 - a) \leq (1 + a)$ in $\mathcal{A}$).
2. $\mathcal{A} \models (\neg\, \phi)(\underset{\sim}{a})$ if and only if not $\mathcal{A} \models \phi(a)$.
3. $\mathcal{A} \models (\phi \;\&\; \psi)(\underset{\sim}{a})$ if and only if $\mathcal{A} \models \phi(\underset{\sim}{a})$ and $\mathcal{A} \models \psi(\underset{\sim}{a})$.
4. $\mathcal{A} \models (\phi \text{ or } \psi)(\underset{\sim}{a})$ if and only if $\mathcal{A} \models \phi(\underset{\sim}{a})$ or $\mathcal{A} \models \psi(\underset{\sim}{a})$.
5. $\mathcal{A} \models (\exists \underset{\sim}{y}\; \phi(\underset{\sim}{x},\underset{\sim}{y}))(\underset{\sim}{a})$ if and only if $\mathcal{A} \models \phi(\underset{\sim}{a},b)$ for some $b \in \mathcal{A}$.

$\mathcal{A}$ is a *model* for a theory T if $\mathcal{A} \models \tau$ for every sentence $\tau \in T$. In this case, we write $\mathcal{A} \models T$. So if T is the set of axioms for totally ordered fields, $\mathcal{A} \models T$ means $\mathcal{A}$ is a totally ordered field. K. Gödel has proved that every theory T has a model.

If $\mathcal{L}$ is a language and $\mathcal{A}$ is a structure for $\mathcal{L}$, $\mathcal{L}(\mathcal{A})$ is the language obtained by adding a constant *a* to $\mathcal{L}$ for each $a \in \mathcal{A}$. (We will usually forget to write *a* and just use a.) The structures of $\mathcal{L}(\mathcal{A})$ will be denoted by $(\mathcal{B},a)_{a\in A}$ where $\mathcal{B}$ is a structure for $\mathcal{L}$. We write $\mathcal{B} \equiv \mathcal{C}$ if for every sentence ϕ of $\mathcal{L}$, $\mathcal{B} \models \phi$ if and only if $\mathcal{C} \models \phi$ and say $\mathcal{B}$ and $\mathcal{C}$ are *elementarily equivalent*. We write $\mathcal{A} \prec \mathcal{B}$ if $\mathcal{A} \subseteq \mathcal{B}$ and $(\mathcal{A},a)_{a\in A} \equiv (\mathcal{B},a)_{a\in A}$ (i.e., $\mathcal{A}$ and $\mathcal{B}$ are elementarily equivalent in $\mathcal{L}(\mathcal{A})$) and say $\mathcal{A}$ is an *elementary submodel* of $\mathcal{B}$.

Often a theory can be augmented to obtain a nicer theory. For example, if T is the theory of fields of characteristic 0 (i.e., T is the usual finite list of axioms for fields together with the infinite list of axioms (p), one for each prime p; (p) $\forall x\, (\neg\; px = 0 \text{ or } x = 0)$ where px is a shorthand for the sum of p x's) we can add axioms saying every polynomial has a solution to obtain algebraically closed fields of characteristic 0. This is the model companion. In general, for a theory T, T^c is a *model companion* of T if

1. Every model of T can be embedded in a model of T^c, and conversely, and
2. T^c is *model-complete* (If $\mathcal{A} \subseteq \mathcal{B}$ are models of T^c, then $\mathcal{A} \prec \mathcal{B}$).

If, in addition, every model of T^c is a model of T and whenever $\mathcal{A} \models T$, $\mathcal{B}$, $\mathcal{C} \models T^c$ and $\mathcal{A} \subseteq \mathcal{B}$, $\mathcal{C}$, then $(\mathcal{B},a)_{a\in A} \equiv (\mathcal{C},a)_{a\in A}$, we say T^c is the *model completion* of T.

The following lemma simplifies companionability proofs. The first part is due to A. Robinson and the second is model-theoretic folklore, easily proved:

LEMMA 0. (A. Robinson) A theory T is model-complete if and only if, whenever $\mathcal{A} \subseteq \mathcal{B}$ are models of T and $\phi(\underset{\sim}{x})$ is a quantifier-free formula of $\mathcal{L}(\mathcal{A})$

such that $\mathcal{B} \models \exists \underset{\sim}{x}\phi(\underset{\sim}{x})$, then $\mathcal{A} \models \exists \underset{\sim}{x}\phi(\underset{\sim}{x})$. Furthermore, if T is a universal theory (i.e., axiomatizable by formulae of the form $\forall \underset{\sim}{x}\psi(\underset{\sim}{x})$, $\psi(\underset{\sim}{x})$ quantifier-free) then it is sufficient to consider formulae $\exists x\phi(x)$ in one free variable x.

If the models of T are closed under unions, we say that T is *inductive*. In this case, the first extra clause for model completion holds. For simplicity, we assume that all theories are inductive unless it is stated to the contrary.

THEOREM 1. Assume T has a model companion T^c. Then T^c is the model completion if and only if T enjoys the amalgamation property (AP).

Proof. If T has AP, assume $\mathcal{A} \models T$, $\mathcal{B},\mathcal{C} \models T^c$ and $\mathcal{A} \subseteq \mathcal{B},\mathcal{C}$. There exists $\mathcal{B}',\mathcal{C}' \models T^c$ such that $\mathcal{B}' \supseteq \mathcal{B}$ and $\mathcal{C}' \supseteq \mathcal{C}$. By AP, the diagram

$$\begin{array}{ccc} \mathcal{A} & \longrightarrow & \mathcal{B}' \\ \downarrow & & \downarrow \\ \mathcal{C}' & \dashrightarrow & \mathcal{D} \models T \end{array}$$

can be completed so all indicated maps are one-to-one. There is $\mathcal{D}^* \models T^c$ such that $\mathcal{D}^* \supseteq \mathcal{D}$ (by 1). Now $\mathcal{B}$, $\mathcal{C}$, $\mathcal{D}^* \models T^c$ so $\mathcal{B} \prec \mathcal{D}^*$ and $\mathcal{C} \prec \mathcal{D}^*$. Hence, $(\mathcal{B},a)_{a\in A} \equiv (\mathcal{D}^*,a)_{a\in A} \equiv (\mathcal{C},a)_{a\in A}$ and T^c is the model completion of T.

Conversely, if T^c is the model completion of T, let $\mathcal{A} \subseteq \mathcal{B}$, $\mathcal{C}$ where $\mathcal{A}$, $\mathcal{B}$, $\mathcal{C} \models T$. There exist $\mathcal{B}^*$, $\mathcal{C}^* \models T^c$ such that $\mathcal{B}^* \supseteq \mathcal{B}$ and $\mathcal{C}^* \supseteq \mathcal{C}$. By hypothesis $(\mathcal{B}^*, a)_{a\in A} \equiv (\mathcal{C}^*, a)_{a\in A}$, so $T^c \cup D(\mathcal{B}^*) \cup D(\mathcal{C}^*)$ is consistent ($D(\mathcal{B}^*)$, the *diagram* of $\mathcal{B}^*$, is the set of all atomic and negated atomic sentences of $\mathcal{L}(\mathcal{B}^*)$ which hold in $\mathcal{B}^*$) and so has a model $\mathcal{D}^*$. Since $\mathcal{D}^* \models T^c$ then it is embeddable in a model $\mathcal{D}$ of T, which is the desired amalgamation.

From now on, we use the phrase *basic sentence* for any sentence that is either atomic or negated atomic.

Let $\mathcal{A}$, $\mathcal{B}$ be structures of $\mathcal{L}$ with $\mathcal{A} \subseteq \mathcal{B}$. $\mathcal{A}$ is *existentially complete* in $\mathcal{B}$ if whenever $\phi(\underset{\sim}{x}, \underset{\sim}{a})$ is a quantifier-free formula in $\mathcal{L}(\mathcal{A})$ and $\underset{\sim}{a} \in \mathcal{A}$, $\mathcal{B} \models \exists \underset{\sim}{x}\phi(\underset{\sim}{x},\underset{\sim}{a})$ implies $\mathcal{A} \models \exists \underset{\sim}{x}\phi(\underset{\sim}{x},\underset{\sim}{a})$. If ϕ is restricted to a conjunction of atomic formulae, we say that $\mathcal{A}$ is *algebraically closed* in $\mathcal{B}$. For example, the group $\underset{\sim}{Q}$ of rationals is algebraically closed in $\mathcal{B} = \underset{\sim}{Q} \oplus \mathbb{Z}_2$ (divisibility is sufficient for this) but not existentially closed: the sentence $\exists x(x + x = 0$ and $x \neq 0)$ is satisfiable in $\mathcal{B}$ but not Q. For fields, the two concepts coincide since the solvability of $p(\underset{\sim}{x}) \neq q(\underset{\sim}{x})$ can be replaced by that of $y(p(\underset{\sim}{x}) - q(\underset{\sim}{x})) = 1$.

Let $\mathcal{A}$ be a structure for $\mathcal{L}$, and $\mathcal{U}$ an ultrafilter on a set I. Define a relation $\sim$ on $\prod_{i\in I} \mathcal{A}$ by: $f \sim g$ if $\{i \in I\colon\ f(i) = g(i)\} \in \mathcal{U}$ (intuitively, f and g agree almost everywhere). $\sim$ is a congruence relation, so the set of congruence classes (called an *ultrapower* of $\mathcal{A}$ and denoted $\Pi\mathcal{A}/\mathcal{U}$) is a structure for $\mathcal{L}$ in the natural way (e.g., $f^{\sim} < g^{\sim}$ if $f(i) < g(i)$ almost everywhere).

ŁOS' THEOREM. Let $\mathcal{A}$ be a structure for $\mathcal{L}$. Then $\mathcal{A} \equiv \Pi\mathcal{A}/\mathcal{U}$ and $\mathcal{A} \prec \Pi\mathcal{A}/\mathcal{U}$ via the natural map $a \mapsto (\ldots,a,\ldots)^{\sim} = a^{\sim}$. Actually, $\mathcal{A} \models \phi(\underset{\sim}{a})$ if and only if $\Pi\mathcal{A}/\mathcal{U} \models \phi(\underset{\sim}{a}^{\sim})$.

The proof is by induction on the complexity of formulae ϕ.

LEMMA 2. Assume $\mathcal{L}$ has only a finite number of constant, operation and function symbols. Suppose $\mathcal{A} \subseteq \mathcal{B}$. Then $\mathcal{A}$ is algebraically closed (existentially complete) in $\mathcal{B}$ if and only if there is an ultrafilter $\mathcal{U}$ such that for some homo (mono) morphism η, the following diagram commutes.

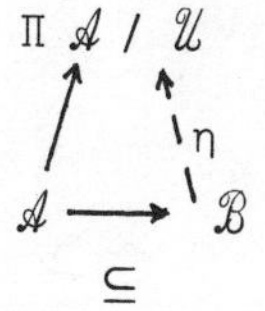

Proof. If such a diagram exists and $\mathcal{B} \models \exists \underset{\sim}{x}\phi(\underset{\sim}{x},\underset{\sim}{a})$, then $\mathcal{A} \models \phi(\underset{\sim}{b},\underset{\sim}{a})$ for some $\underset{\sim}{b} \in \mathcal{B}$. Hence $\Pi\mathcal{A}/\mathcal{U} \models \phi(\underset{\sim}{b}\eta,\underset{\sim}{a}\eta)$ (for algebraically closed, ϕ has no negations so is satisfied by homomorphic images). So $\mathcal{A} \models \exists \underset{\sim}{x}\phi(\underset{\sim}{x},\underset{\sim}{a})$ by Łos' Theorem. Thus $\mathcal{A}$ is algebraically closed in $\mathcal{B}$ (or existentially complete in $\mathcal{B}$ if η is monomorphism).

Conversely, replace all operation symbols in $\mathcal{L}$ by relation symbols (e.g., $x + y$ by $S(x,y,z)$, standing for $x + y = z$). Let I be the set of all pairs $i = (S_1,S_2)$ where $S_1 \subseteq \mathcal{A}$, $S_2 \subseteq \mathcal{B} \setminus \mathcal{A}$ and both are finite. Say, $S_2 = \{b_1,\ldots,b_n\}$. Let $\mathcal{S}$ be the set of equations (and inequations) in $\underset{\sim}{x}$ with constants from S_1 which are satisfied in $\mathcal{B}$ by $\underset{\sim}{b}$. Since $\mathcal{S}$ is finite, it has a solution $\underset{\sim}{a}$ in $\mathcal{A}$. Let $\mathcal{B}_i \subseteq \mathcal{B}$ have universe $S_1 \cup S_2 \cup \{$constants of $\mathcal{L}\}$ (so $\mathcal{B}_i$ is finite). Then there is a homo (mono) morphism $\eta_i\colon\ \mathcal{B}_i \to \mathcal{A}$ which is the identity except on S_2, where we define $b_j\eta_i = a_j$ $(1 \le j \le n)$. Let $k \in I$, and $X_k = \{i \in I\colon\ \mathcal{B}_k \subseteq \mathcal{B}_i\}$. Then $\{X_k\colon\ k \in I\}$ has the finite intersection property so is contained in an ultrafilter $\mathcal{U}$. Let $\eta\colon\ \mathcal{B} \to \Pi\mathcal{A}/\mathcal{U}$ be defined via $(b\eta)_i =$

$b\eta_i$ if $b \in \mathcal{B}_i$, and arbitrary otherwise. Then η is the required homo (mono) morphism.

Let $\mathcal{K}$ be a class of structures of $\mathcal{L}$ and $\mathcal{S}(\mathcal{K})$ be the substructures of members of $\mathcal{K}$. $\mathcal{A} \in \mathcal{S}(\mathcal{K})$ is $\mathcal{K}$-existentially complete ($\mathcal{K}$-algebraically closed) if $\mathcal{A}$ is existentially complete (algebraically closed) in every member of $\mathcal{K}$ that contains it. If $\mathcal{K} = \text{Mod}(T)$ for some theory T, we will simply replace $\mathcal{K}$ by T in the above. If T can be axiomatised by a set of universal sentences then $\mathcal{S}(T) = T$, and therefore all T-existentially complete structures are in fact models of T. Most algebraic theories fall into this classification; e.g., semigroups, (abelian)groups, (abelian)o-groups, (abelian)ℓ-groups, rings (if we allow the additive inverse as a unary operation in $\mathcal{L}$).

$\mathcal{A} \in \mathcal{K}$ is *simple* in $\mathcal{K}$ if whenever $\eta:\mathcal{A} \to \mathcal{B}$ is a homomorphism, then η is constant or a monomorphism.

THEOREM 3. (P. Bacsich) Let $\mathcal{K}$ be a class of structures of a finite language $\mathcal{L}$ that is closed under ultrapowers and has the property that every member of $\mathcal{K}$ is contained in a simple member of $\mathcal{K}$. Then any $\mathcal{K}$-algebraically closed structure of cardinality greater than 1 is $\mathcal{K}$-existentially complete.

Proof. Let $|\mathcal{A}| > 1$, $\mathcal{A}$ $\mathcal{K}$-algebraically closed and let $\mathcal{A} \subseteq \mathcal{B} \in \mathcal{K}$. There is $\mathcal{C} \in \mathcal{K}$ such that $\mathcal{C}$ is simple. By Lemma 2, there is a homomorphism $\eta:\mathcal{C} \to \Pi\mathcal{A}/\mathcal{U}$ for some ultrafilter $\mathcal{U}$. Since $\eta|\mathcal{A}$ is the canonical map, η is not constant. Since $\mathcal{C}$ is simple, η is a monomorphism, so the same is true of $\eta|\mathcal{B}$. Hence $\mathcal{A}$ is existentially complete in $\mathcal{B}$ by Lemma 2. This is for all $\mathcal{B} \in \mathcal{K}$ with $\mathcal{A} \subseteq \mathcal{B}$.

The presentation given here can be found in [5].

COROLLARY 4. For the following theories, all algebraically closed structures of cardinality greater than 1 are existentially complete: (a) groups (b) o-groups (c) ℓ-groups.

After this aside, let's return to model companions.

LEMMA 5. Every model of T is contained in an existentially complete one.

Proof. Let $\mathcal{A} \models T$. Form $\mathcal{A}^\dagger$ as follows: enumerate the existential sentences of $\mathcal{L}(\mathcal{A})$, $\phi_0, \phi_1, \phi_2, \ldots, \phi_\lambda \ldots$. Form $\mathcal{A} = \mathcal{A}_0 \subseteq \mathcal{A}_1 \subseteq \cdots \subseteq \mathcal{A}_\lambda \subseteq \cdots$ such that: if ϕ_λ is satisfiable in some $\mathcal{B} \supseteq \mathcal{A}_\lambda$ with $\mathcal{B} \models T$, let $\mathcal{A}_{\lambda+1}$ be the submodel of $\mathcal{B}$ generated by $\mathcal{A}_\lambda$ and a(finite)solution to ϕ_λ in $\mathcal{B}$; otherwise let $\mathcal{A}_{\lambda+1} = \mathcal{A}_\lambda$. If λ is a limit ordinal, let $\mathcal{A}_\lambda = \bigcup_{\mu<\lambda} \mathcal{A}_\mu$. Let $\mathcal{A}^\dagger = \bigcup \mathcal{A}_\lambda$. Note $\mathcal{A}^\dagger \in \mathcal{S}(T)$. Now if ϕ is an existential sentence in $\mathcal{L}(\mathcal{A})$ that holds in some $\mathcal{B} \models T$ with $\mathcal{B} \supseteq \mathcal{A}^\dagger$ then it holds in $\mathcal{A}^\dagger$. Finally, let $\mathcal{A}(0) = \mathcal{A}$, $\mathcal{A}(n + 1) = \mathcal{A}(n)^\dagger$ and $\mathcal{C} = \bigcup_{n\in\omega} \mathcal{A}(n)$; $\mathcal{C}$ is clearly the desired T-existentially complete model.

Observe that $|\mathcal{C}| = \max\{|\mathcal{A}|, \aleph_0\}$.

It follows that if T^c exists, its models are precisely the T-existentially complete structures (therefore a theory has at most one model companion). Hence, by Łos' theorem, to show that T^c doesn't exist, we must find a T-existentially complete structure $\mathcal{A}$ with $\Pi\mathcal{A}/\mathcal{U}$ not T-existentially complete in T for some ultrafilter $\mathcal{U}$. This leads us to trying to find existentially complete $\mathcal{A}$, $\mathcal{B}$ which are not elementarily equivalent. There are two ways to build existentially complete structures, by finite and by infinite forcing. The idea is this. Let C be a set of constant symbols not occurring in $\mathcal{L}$. Let ϕ be a formula in a language $\mathcal{L}' = \mathcal{L}(C) \supseteq \mathcal{L}$, and let A be a condition (we'll say what this is soon). We'll declare which atomic sentences are forced by A ($A \# \phi$) and define $A \# \phi$ (A *forces* ϕ) inductively for more complicated ϕ by:

$A \# \phi \,\&\, \psi \leftrightarrow A \# \phi$ and $A \# \psi$

$A \# \phi$ or $\psi \leftrightarrow A \# \phi$ or $A \# \psi$

$A \# \exists x\phi(x) \leftrightarrow A \# \phi(c)$ for some $c \in C$.

(*) $A \# \neg\phi \leftrightarrow$ For no condition $B \supseteq A$, $B \# \phi$.

Note that with the exception of negation, this parallels the $\models$ relation.

For *infinite forcing* ($\Vdash$) conditions A will be members $\mathcal{B} \supseteq \mathcal{A}$ of $\mathcal{S}(T)$, $C = \mathcal{A}$ (so $\mathcal{L}(C) = \mathcal{L}(\mathcal{A})$) and $\mathcal{A} \Vdash \phi$ for ϕ atomic if and only if $\mathcal{A} \models \phi$. So we're considering $\mathcal{A}$ forcing sentences of the language $\mathcal{L}(\mathcal{A})$, $\mathcal{A}$ ranging through $\mathcal{S}(T)$.

For *finite forcing* ($\Vdash$) C is countable, conditions will be finite sets, p, of basic sentences of $\mathcal{L}(C)$ consistent with T, and $p \Vdash \phi$ for ϕ atomic if and only if $\phi \in p$.

Finite forcing only looks at a finite amount of information at a time and so requires some finesse. Infinite forcing is comparatively crude. This should be borne in mind for Theorems 7 and 19.

LEMMA 6. No condition forces ϕ and $\neg\phi$. If $A \# \phi$ and $B \supseteq A$ is a condition then $B \# \phi$, where $\#$ is $\Vdash$ or $\Vdash$.

The first half is immediate from the definition. The second half is proved by induction on ϕ.

$\mathcal{A} \in \mathcal{S}(T)$ is *infinitely generic* if, for every sentence ϕ of $\mathcal{L}(\mathcal{A})$, $\mathcal{A} \Vdash \phi$ or $\mathcal{A} \Vdash \neg\phi$.

THEOREM 7. If $\mathcal{A} \models T$, there is an infinitely generic $\mathcal{B} \supseteq \mathcal{A}$.

Proof. Similar to the proof of Lemma 5. Enumerate the sentences of $\mathcal{L}(\mathcal{A})$, $\phi_0, \phi_1, \phi_2, \ldots, \phi_\lambda, \ldots$. Define a sequence $\mathcal{A} = \mathcal{A}_0 \subseteq \mathcal{A}_1 \subseteq \mathcal{A}_2 \subseteq \cdots \subseteq \mathcal{A}_\lambda \subseteq \cdots$ of models of T recursively as follows: if λ is a limit ordinal, let $\mathcal{A}_\lambda = \bigcup_{\mu<\lambda} \mathcal{A}_\lambda$ Let $\lambda = \mu + 1$. If $\mathcal{A}_\mu \Vdash \phi_\mu$ or $\mathcal{A}_\mu \Vdash \neg\phi_\mu$ let $\mathcal{A}_\lambda = \mathcal{A}_\mu$. Otherwise, for some $\mathcal{B}' \models T$, $\mathcal{B}' \supseteq \mathcal{A}$, $\mathcal{B}' \Vdash \phi_\mu$ (since not $\mathcal{A}_\mu \Vdash \neg\phi_\mu$). Let $\mathcal{A}_\lambda$ be such a $\mathcal{B}'$. Let $\mathcal{A}^\dagger = \bigcup \mathcal{A}_\lambda$. For each sentence ϕ of $\mathcal{L}(\mathcal{A})$, $\mathcal{A}^\dagger \Vdash \phi$ or $\mathcal{A}^\dagger = \neg\phi$. Let $\mathcal{A}(0) = \mathcal{A}$, $\mathcal{A}(n+1) = \mathcal{A}(n)^\dagger$ and $\mathcal{B} = \bigcup_{n\in\omega} \mathcal{A}(n)$. It is straightforward to see $\mathcal{B}$ is infinitely generic (for if ϕ is a sentence in $\mathcal{L}(\mathcal{A})$, it is a sentence in $\mathcal{L}(\mathcal{A}(n))$ for some n and so $\mathcal{A}(n+1) \Vdash \phi$ or $\mathcal{A}(n+1) \Vdash \neg\phi$).

An easy induction on the complexity of ϕ gives

PROPOSITION 8. If $\mathcal{A}$ is infinitely generic and ϕ is a sentence of $\mathcal{L}(\mathcal{A})$, $\mathcal{A} \Vdash \phi$ if and only if $\mathcal{A} \models \phi$.

COROLLARY 9. If $\mathcal{A}$, $\mathcal{B}$ are infinitely generic and $\mathcal{A} \subseteq \mathcal{B}$, then $\mathcal{A} \prec \mathcal{B}$.

Proof. If ϕ is a sentence of $\mathcal{L}(\mathcal{A})$ and $\mathcal{A} \models \phi$ but $\mathcal{B} \models \neg\phi$, then $\mathcal{A} \Vdash \phi$ but $\mathcal{B} \Vdash \neg\phi$. This contradicts Lemma 6.

LEMMA 10. If $\mathcal{A}$ is infinitely generic, then it is existentially complete in T.

Proof. Let ϕ be an existential sentence of $\mathcal{L}(\mathcal{A})$ which is true in some $\mathcal{B} \supseteq \mathcal{A}$ with $\mathcal{B} \models T$. By Theorem 7, there is an infinitely generic $\mathcal{D} \supseteq \mathcal{B}$. Since $\mathcal{B} \models \phi$ and ϕ is existential, $\mathcal{D} \models \phi$. By Corollary 9, $\mathcal{A} \prec \mathcal{D}$ (since $\mathcal{A} \subseteq \mathcal{B} \subseteq \mathcal{D}$) so $\mathcal{A} \models \phi$. Thus $\mathcal{A}$ is existentially complete in T.

T has the *joint embedding property* if whenever $\mathcal{A}, \mathcal{B} \models T$, there is $\mathcal{D} \models T$ such that $\mathcal{A}, \mathcal{B} \subseteq \mathcal{D}$. e.g., if T is the theory of ℓ-groups, $\mathcal{D}$ can be taken as the direct sum of $\mathcal{A}$ and $\mathcal{B}$.

LEMMA 11. If T has the joint embedding property, then $\mathcal{A} \equiv \mathcal{B}$ for any $\mathcal{A}$, $\mathcal{B}$ infinitely generic.

Proof. Let ϕ be a sentence of $\mathcal{L}$ and $\mathcal{A}$, $\mathcal{B}$ be infinitely generic. Then there exist $\mathcal{A}'$, $\mathcal{B}' \models T$ such that $\mathcal{A} \subseteq \mathcal{A}'$ and $\mathcal{B} \subseteq \mathcal{B}'$ ($\mathcal{A}$, $\mathcal{B} \in \mathcal{S}(T)$). Hence there exists $\mathcal{C}' \models T$ such that $\mathcal{A}', \mathcal{B}' \subseteq \mathcal{C}'$. By Theorem 7, there exists an infinitely generic $\mathcal{C} \supseteq \mathcal{C}'$. Now $\mathcal{A}$, $\mathcal{B} \subseteq \mathcal{C}$ so, by Corollary 9, $\mathcal{A} \prec \mathcal{C}$ and $\mathcal{B} \prec \mathcal{C}$. Hence $\mathcal{A} \equiv \mathcal{C} \equiv \mathcal{B}$.

The converse of Lemma 11 is also true.

This completes our discussion of infinitely generic structures. As we have seen, this class consists of "large" existentially complete structures, all of which are elementarily equivalent.

Before embarking on finitely generic ones, we provide some examples of forcing to familiarize the reader with the definition and technique.

EXAMPLE 1. Let T be the theory of linear ordering. Then $\emptyset \Vdash \forall x \exists y(x < y)$ (i.e., no largest element) i.e., $\emptyset \Vdash \neg \exists x \neg \exists y(x < y)$. Suppose not. Then there is a condition p such that $p \Vdash \exists x \neg \exists y(x < y)$; i.e., $p \Vdash \neg \exists y(c < y)$ for some $c \in C$. Now p is a finite set. Let $c_0 \in C$ be a constant not occurring in p, and let $q = p \cup \{c < c_0\}$. Then $q \cup T$ is consistent so q is a condition containing p. Now $q \Vdash c < c_0$ since $(c < c_0) \in q$. Hence $q \Vdash \exists y(c < y)$. This contradicts $p \Vdash \neg \exists y(c < y)$. Therefore $\emptyset \Vdash \neg \exists x \neg \exists y(x < y)$.

EXAMPLE 2. For the same theory, $\emptyset \Vdash \forall x \forall y \exists z(x < z < y \text{ or } x \nless y)$(i.e., denseness). Exercise for the reader.

EXAMPLE 3. Let T be the theory of groups. Then $\emptyset \Vdash \neg\neg \exists x \exists y\,(xy \neq yx)$ i.e., for any p, there is $q \supseteq p$ such that $q \Vdash \exists x \exists y\,(xy \neq yx)$. Let p be any condition and $c_0, c_1 \in C$ not occur in p. For some group G, p is true in G ($p \subseteq D(G)$ under some interpretation of the constants in p). Now if H is any non-abelian group, $G \oplus H$ satisfies $p \cup \{c_0c_1 \neq c_1c_0\} = q$. So q is a condition, $q \supseteq p$, and $q \Vdash c_0c_1 \neq c_1c_0$, since $(c_0c_1 \neq c_1c_0) \in q$. Hence $q \Vdash \exists x \exists y\,(xy \neq yx)$.

Note that $\emptyset$ forces $\neg\neg \exists x \exists y\,(xy \neq yx)$ but doesn't force $\exists x \exists y (xy \neq yx)$ even though the two sentences are equivalent.

The following lemma is straightforward and will be used implicitly in future proofs.

LEMMA 12. (i) $p \Vdash \phi$ implies $p \Vdash \neg\neg\phi$ (ii) $p \Vdash \neg\phi$ if and only if $p \Vdash \neg\neg\neg\,\phi$. (iii) If ϕ is basic and $p \Vdash \phi$, then $p \cup \{\phi\}$ is a condition (iv) If ϕ is universal, then $p \Vdash \phi$, if and only if ϕ is deducible from $T \cup p$.

Denote by T^f the set of all sentences ϕ of $\mathcal{L}$ such that $\emptyset \Vdash \neg\neg\,\phi$. T^f is called the *finite forcing companion of* T.

Let $\mathcal{A} \in \mathcal{S}(T)$. Write $\mathcal{A} \Vdash \phi$ for ϕ a sentence of $\mathcal{L}(\mathcal{A})$ if there is a finite $p \subseteq D(\mathcal{A})$ such that $p \Vdash \phi$. (Since $p \subseteq D(\mathcal{A})$ and $\mathcal{A} \in \mathcal{S}(T)$, $p \cup T$ is consistent.)

$\mathcal{A} \in \mathcal{S}(T)$ is *finitely generic* if for every sentence ϕ of $\mathcal{L}(\mathcal{A})$, $\mathcal{A} \models \phi$ if and only if $\mathcal{A} \Vdash \phi$.

LEMMA 13. If $\mathcal{A}$ is finitely generic, then $\mathcal{A} \models T^f$.

PROOF. Suppose $\phi \in T^f$ and $\mathcal{A} \models \neg\phi$. Then for some $p \subseteq D(\mathcal{A})$, $p \Vdash \neg\,\phi$. Hence $\emptyset \Vdash \neg\neg\,\phi$ fails, a contradiction.

We will prove the converse soon. The next theorem is the analogue of Theorem 7.

THEOREM 14. Let p be a condition relative to T. Then there is a finitely generic $\mathcal{A}$ such that $p \subseteq D(\mathcal{A})$.

Proof. Let C be countable and enumerate the sentences of $\mathcal{L}(C)$: $\phi_0, \phi_1, \phi_2, \ldots$. Let $p_0 = p$. Construct a complete sequence of conditions;

that is $p_0 \subseteq p_1 \subseteq p_2 \subseteq \ldots$ such that for every $n \in \omega$, $p_{n+1} \Vdash \phi_n$ or $p_{n+1} \Vdash \neg \phi_n$. This is straightforward induction: If $p_n \Vdash \neg \phi_n$, let $p_{n+1} = p_n$, or $p_n \cup \{\neg\phi_n\}$ if ϕ_n is atomic. Otherwise there is $q \supseteq p_n$ such that $q \Vdash \phi_n$. Let $p_{n+1} = q \cup \{\phi_n\}$ if ϕ_n basic and $p_{n+1} = p_n$ otherwise. Define $c \sim c'$ if $p_n \Vdash c = c'$ for some $n \in \omega$ $(c, c' \in C)$. This gives an equivalence relation on C. Let $\mathcal{A}$ be the structure formed by the equivalence classes (e.g., $c^{\sim} \leq c'^{\sim}$ (in $\mathcal{A}$) provided $p_n \Vdash c \leq c'$ for some $n \in \omega$ — i.e., $(c \leq c') \in p_n$ for some $n \in \omega$. This is well defined). Since any finite $q \subseteq D(\mathcal{A})$ is contained in some p_n and $p_n \cup T$ is consistent, $q \cup T$ is consistent, so $\mathcal{A} \in \mathcal{S}(T)$. Now if ϕ is atomic then $\phi \in p_n$ or $\neg \phi \in p_n$ for some $n \in \omega$. Thus, if ϕ is atomic, $\mathcal{A} \models \phi$ if and only if $\mathcal{A} \Vdash \phi$. We now show, by induction on ψ, $\mathcal{A} \models \psi$ if and only if $\mathcal{A} \Vdash \psi$. The only difficulty is in passing from ψ to $\neg \psi$. If $\mathcal{A} \Vdash \neg \psi$, not $\mathcal{A} \Vdash \psi$ so not $\mathcal{A} \models \psi$ by induction hypothesis. Hence $\mathcal{A} \models \neg \psi$. Conversely, let $\mathcal{A} \models \neg \psi$. Now ψ is ϕ_m for some m, so $p_{m+1} \Vdash \psi$ or $p_{m+1} \Vdash \neg \psi$. The former would yield $\mathcal{A} \models \psi$ by induction. Hence $p_{m+1} \Vdash \neg \psi$, so $\mathcal{A} \Vdash \neg \psi$.

The last part of the proof gives:

COROLLARY 15. $\mathcal{A} \in \mathcal{S}(T)$ is finitely generic if and only if for every sentence ϕ of $\mathcal{L}(\mathcal{A})$, $\mathcal{A} \Vdash \phi$ or $\mathcal{A} \Vdash \neg \phi$.

COROLLARY 16 (cf Corollary 9). If $\mathcal{A}$, $\mathcal{B}$ are finitely generic and $\mathcal{A} \subseteq \mathcal{B}$, then $\mathcal{A} \prec \mathcal{B}$.

Proof. Let ϕ be a sentence of $\mathcal{L}(\mathcal{A})$ such that $\mathcal{A} \models \phi$. Then $\mathcal{A} \Vdash \phi$ so for some finite $p \subseteq D(\mathcal{A})$, $p \Vdash \phi$. $p \subseteq D(\mathcal{A}) \subseteq D(\mathcal{B})$ so $\mathcal{B} \Vdash \phi$ and hence $\mathcal{B} \models \phi$.

COROLLARY 17 (converse of Lemma 13). If $\phi \notin T^f$, there is a finitely generic $\mathcal{A}$ such that $\mathcal{A} \models \neg \phi$.

Proof. If $\phi \notin T^f$, some condition $p \Vdash \neg \phi$. By Theorem 14, there is a finitely generic $\mathcal{A}$ such that $p \subseteq D(\mathcal{A})$. Since $\mathcal{A} \Vdash \neg \phi$, $\mathcal{A} \models \neg \phi$.

LEMMA 18 (cf Lemma 10). If $\mathcal{A}$ is finitely generic, then it is existentially complete in T.

Proof. Let $\mathcal{B} \models T$ and $\mathcal{A} \subseteq \mathcal{B} \models \exists \underset{\sim}{x}\phi(\underset{\sim}{x},\underset{\sim}{a})$, $\underset{\sim}{a} \in \mathcal{A}$, ϕ a quantifier-free formula of $\mathcal{L}(\mathcal{A})$. If $\mathcal{A} \models \neg \exists \underset{\sim}{x}\phi(\underset{\sim}{x},\underset{\sim}{a})$, then there is finite $p \subseteq D(\mathcal{A}) \subseteq D(\mathcal{B})$ with $p \Vdash \neg \exists \underset{\sim}{x}\phi(\underset{\sim}{x},\underset{\sim}{a})$. But $\mathcal{B} \models \phi(\underset{\sim}{b},\underset{\sim}{a})$ for some $\underset{\sim}{b} \in \mathcal{B}$. Let q be the set of all basic formulae occurring in ϕ which are true in $\mathcal{B}$. Then $q \subseteq D(\mathcal{B})$ and so is a condition, and clearly $q \Vdash \exists \underset{\sim}{x}\phi(\underset{\sim}{x},\underset{\sim}{a})$. But $p \cup q \subseteq D(\mathcal{B})$ and so is a condition containing p and q. Thus it forces $\neg\exists \underset{\sim}{x}\phi(\underset{\sim}{x},\underset{\sim}{a})$ and $\exists \underset{\sim}{x}\phi(\underset{\sim}{x},\underset{\sim}{a})$, a contradiction.

THEOREM 19. If $\mathcal{B}$ is finitely generic, $\mathcal{A}$ is existentially complete and $\mathcal{A} \subseteq \mathcal{B}$, then $\mathcal{A}$ is finitely generic.

Proof. We first show $\mathcal{A} \Vdash \phi$ if and only if $\mathcal{B} \Vdash \phi$, ϕ a sentence $\mathcal{L}(\mathcal{A})$. Clearly, $\mathcal{A} \Vdash \phi$ implies $\mathcal{B} \Vdash \phi$. If $\mathcal{B} \Vdash \phi$, there is finite $p \subseteq D(\mathcal{B})$ with $p \Vdash \phi$. Let $b_0,\ldots,b_{n-1}$ be the elements of $\mathcal{B}$ occurring in p which do not belong to $\mathcal{A}$. Then $\mathcal{B} \models \exists \underset{\sim}{x} \bigwedge p(\underset{\sim}{x})$ where $p(\underset{\sim}{x})$ is obtained by replacing b_i by x_i. This is an existential sentence in $\mathcal{L}(\mathcal{A})$, so $\mathcal{A} \models \exists \underset{\sim}{x} \bigwedge p(\underset{\sim}{x})$ since $\mathcal{A}$ is existentially complete in T. Let $a_0,\ldots,a_{n-1} \in \mathcal{A}$ such that $\mathcal{A} \models \bigwedge p(\underset{\sim}{a})$. Then $p(\underset{\sim}{a}) \subseteq D(\mathcal{A})$ and $p(\underset{\sim}{a}) \Vdash \phi$. Thus $\mathcal{A} \Vdash \phi$, completing the proof of the claim.

We prove $\mathcal{A} \models \phi$ if and only if $\mathcal{A} \Vdash \phi$ by induction on ϕ. The only difficulty is in passing from ϕ to $\neg\phi$. If $\mathcal{A} \Vdash \neg\phi$, then $\mathcal{A} \Vdash \phi$ is false so $\mathcal{A} \models \phi$ is false (by induction). Hence $\mathcal{A} \models \neg\phi$. Conversely, if $\mathcal{A} \models \neg\phi$, then not $\mathcal{A} \models \phi$, so $\mathcal{A} \Vdash \phi$ is false. Hence $\mathcal{B} \Vdash \phi$ is false by the claim. By Corollary 15, $\mathcal{B} \Vdash \neg\phi$ so, again by the claim, $\mathcal{A} \Vdash \neg\phi$.

This shows that the finitely generic structures are low down in the class of existentially complete structures. We note without proof (and without precision)

LEMMA 20. T^f is no more complicated than first order number theory.

LEMMA 21. (cf Lemma 11) If T enjoys the joint embedding property and $\mathcal{A}$, $\mathcal{B}$ are finitely generic, then $\mathcal{A} \equiv \mathcal{B}$.

Proof. Let $\mathcal{A}$, $\mathcal{B}$ be finitely generic and let ϕ be a sentence of $\mathcal{L}$. Let $\mathcal{A}', \mathcal{B}' \models T$ with $\mathcal{A}' \supseteq \mathcal{A}$ and $\mathcal{B}' \supseteq \mathcal{B}$. By the joint embedding property there is $\mathcal{C} \models T$ with $\mathcal{C} \supseteq \mathcal{A}', \mathcal{B}'$. There are finite p, q contained in $D(\mathcal{A})$ and $D(\mathcal{B})$ such that ($p \Vdash \phi$ or $p \Vdash \neg\phi$) and ($q \Vdash \phi$ or $q \Vdash \neg\phi$). Since $p \cup q \subseteq D(\mathcal{C})$, we must have ($p \Vdash \phi$ and $q \Vdash \phi$) or ($p \Vdash \neg\phi$ and $q \Vdash \neg\phi$); i.e. $\mathcal{A} \Vdash \phi$ if and only if $\mathcal{B} \Vdash \phi$. Hence $\mathcal{A} \models \phi$ if and only if $\mathcal{B} \models \phi$.

The converse is also true.

PROPOSITION 22. Let T enjoy the joint embedding property.

1. Any $\exists$ sentence which holds in some model of T holds in all T-existentially complete structures.

2. Any $\forall\exists$ sentence which holds in all models of T holds in all T-existentially complete structures.

3. Finitely and infinitely generic structures satisfy the same $\forall\exists$ sentences.

Proof. 1. Let $\mathcal{A}$ be a model of T, $\mathcal{A} \models \exists \underset{\sim}{x}\phi(\underset{\sim}{x})$ where $\phi(\underset{\sim}{x})$ is quantifier-free and let $\mathcal{B}$ be T-existentially complete. By definition, there is $\mathcal{B}' \supseteq \mathcal{B}$ such that $\mathcal{B}'$ is a model of T. Let $\mathcal{C}$ be a model of T containing $\mathcal{A}$ and $\mathcal{B}'$. Then $\mathcal{C} \models \exists \underset{\sim}{x}\phi(\underset{\sim}{x})$, and as $\mathcal{B}$ is existentially complete in T (and so in $\mathcal{C}$), $\mathcal{B} \models \exists \underset{\sim}{x}\phi(\underset{\sim}{x})$.

2. Let $\forall \underset{\sim}{x} \exists \underset{\sim}{y}\phi(\underset{\sim}{x},\underset{\sim}{y})$ hold in all models of T where $\phi(\underset{\sim}{x},\underset{\sim}{y})$ is quantifier-free. Let $\mathcal{A}$ be existentially complete in T. There is $\mathcal{B} \supseteq \mathcal{A}$ such that $\mathcal{B}$ is a model of T. Now $\mathcal{B} \models \forall \underset{\sim}{x} \exists \underset{\sim}{y}\phi(\underset{\sim}{x},\underset{\sim}{y})$ so if $\underset{\sim}{a} \in \mathcal{A}$, $\mathcal{B} \models \exists \underset{\sim}{y}\phi(\underset{\sim}{a},\underset{\sim}{y})$. But $\mathcal{A}$ is existentially complete in $\mathcal{B}$, so $\mathcal{A} \models \exists \underset{\sim}{y}\phi(\underset{\sim}{a},\underset{\sim}{y})$. Since $\underset{\sim}{a}$ was arbitrary, $\mathcal{A} \models \forall \underset{\sim}{x} \exists \underset{\sim}{y}\phi(\underset{\sim}{x},\underset{\sim}{y})$.

3. By Lemma 13 and Corollary 17, T^f is the set of sentences which hold in all finitely generic structures; and by Lemma 21, all models of T^f are elementarily equivalent. Since any two infinitely generic structures are elementarily equivalent (by Lemma 11), they all satisfy the same complete theory, T^F. Let $\forall \underset{\sim}{x} \exists \underset{\sim}{y}\phi(\underset{\sim}{x},\underset{\sim}{y}) \in T^F$ where $\phi(\underset{\sim}{x},\underset{\sim}{y})$ is quantifier-free. If $\mathcal{A}$ is finitely generic, it is contained in some model of T, and hence in some infinitely generic structure $\mathcal{B}$ by Lemma 7. Since $\mathcal{B}$ is infinitely generic, $\mathcal{B} \models \forall \underset{\sim}{x} \exists \underset{\sim}{y}\phi(\underset{\sim}{x},\underset{\sim}{y})$ and $\mathcal{B} \subseteq \mathcal{C}$ for some model $\mathcal{C}$ of T. Let $\underset{\sim}{a} \in \mathcal{A}$. Then $\mathcal{B} \models \exists \underset{\sim}{y}\phi(\underset{\sim}{a},\underset{\sim}{y})$ so $\mathcal{C} \models \exists \underset{\sim}{y}\phi(\underset{\sim}{a},\underset{\sim}{y})$. Since $\mathcal{A}$ is existentially complete in T (Lemma 18), $\mathcal{A} \models \exists \underset{\sim}{y}\phi(\underset{\sim}{a},\underset{\sim}{y})$. Thus $\mathcal{A} \models \forall \underset{\sim}{x} \exists \underset{\sim}{y}\phi(\underset{\sim}{x},\underset{\sim}{y})$. Conversely, by Lemma 12(iv), the universal sentences which hold in all models of T are precisely those that hold in all models of T^f. Hence if $\mathcal{A}$ is a substructure of a model of T, $D(\mathcal{A}) \cup T^f$ is consistent; i.e., there is $\mathcal{B} \supseteq \mathcal{A}$ such that $\mathcal{B}$ is a model of T^f. In particular, letting $\mathcal{A}$ be infinitely generic, there is $\mathcal{B} \supseteq \mathcal{A}$ such that $\mathcal{B}$ is a model of T^f. So if $\forall \underset{\sim}{x} \exists \underset{\sim}{y}\phi(\underset{\sim}{x},\underset{\sim}{y}) \in T^f$ where $\phi(\underset{\sim}{x},\underset{\sim}{y})$ is quantifier-free, $\mathcal{B} \models \forall \underset{\sim}{x} \exists \underset{\sim}{y}\phi(\underset{\sim}{x},\underset{\sim}{y})$ and (as before), $\mathcal{A} \models \forall \underset{\sim}{x} \exists \underset{\sim}{y}\phi(\underset{\sim}{x},\underset{\sim}{y})$. Thus $\forall \underset{\sim}{x} \exists \underset{\sim}{y}\phi(\underset{\sim}{x},\underset{\sim}{y}) \in T^F$.

Note that we did not need the joint embedding property to prove 2 nor $(T^F)_{\forall\exists} \subseteq T^f$ in 3, where T^F is the set of sentences which hold in all infinitely generic in T structures.

The last general result we will need is due to Angus Macintyre [12]:

THEOREM 23. Let T be a recursively axiomatizible theory of $\mathcal{L}$, and Δ a maximal consistent set of basic formulae in $x_0,\ldots,x_{m-1}$. If Δ is not recursive, there is a finitely generic $\mathcal{A}$ such that for each $\underset{\sim}{a} \in \mathcal{A}$ there is $\delta \in \Delta$ such that $\mathcal{A} \models \neg\, \delta(\underset{\sim}{a})$. ($\mathcal{A}$ *omits* Δ).

Proof. The proof closely resembles that of Theorem 14. Again let $\phi_0, \phi_1, \phi_2,\ldots$ enumerate the sentences of $\mathcal{L}(C)$. Let $\sigma_0, \sigma_1, \sigma_2,\ldots$ be an enumeration of all m-tuples in C. We construct a complete sequence $p_0 \subseteq p_1 \subseteq p_2 \subseteq \ldots$ of conditions such that (1) $p_{2n} \Vdash \phi_n$ or $p_{2n} \Vdash - \phi_n$ $(n \in \omega)$, and if ϕ_n is atomic, $\phi_n \in p_{2n}$ or $\neg\, \phi_n \in p_{2n}$ and (2) there is a sentence $\theta_n(\sigma_n)$ in $\mathcal{L}(C)$ such that $\theta_n(\underset{\sim}{x}) \in \Delta$ and $p_{2n+1} \Vdash \neg\, \theta_n(\sigma_n)$ or $\neg\, \theta_n(\underset{\sim}{x}) \in \Delta$ and $p_{2n+1} \Vdash \theta_n(\sigma_n)$ Achieving (1) is exactly as in the proof of Theorem 14. To achieve (2) note that $\Gamma = \{\psi(\underset{\sim}{x})\colon \psi$ is basic and $T \cup p_{2n} \Vdash \psi\}$ is recursively enumerable. Let Δ^+ be the set of atomic formulae in Δ, $\Delta^- = \Delta \backslash \Delta^+$, and let Γ^+, Γ^- be defined similarly. Clearly, Γ^+, Γ^- are also recursively enumerable. If $\Delta^+ = \Gamma^+$ and $\Delta^- = \Gamma^-$, then Δ^+ and Δ^- are recursively enumerable. Hence Δ is recursive, a contradiction. Hence there is $\theta_n(\underset{\sim}{x}) \in \Delta^+$ such that for some $q \supseteq p_{2n}$, $q \Vdash \neg\, \theta_n(\sigma_n)$, or $\neg\, \theta_n(\underset{\sim}{x}) \in \Delta^-$ and for some $q \supseteq p_{2n}$, $q \Vdash \theta_n(\sigma_n)$. Take p_{2n+1} to be such a q. As before using the sequence $\{p_n\colon\ n \in \omega\}$ we obtain a finitely generic structure $\mathcal{A}$. Moreover each $\underset{\sim}{a} \in \mathcal{A}$ is the equivalence class of some σ_n. But $p_{2n+1} \Vdash \neg\, \theta_n(\sigma_n)$ if $\theta_n(\underset{\sim}{x}) \in \Delta$ and $p_{2n+1} \Vdash \theta_n(\sigma_n)$ if $\neg\, \theta_n(\underset{\sim}{x}) \in \Delta$. Hence $\mathcal{A} = \theta_n(\sigma_n)$ if and only if $\theta_n(\underset{\sim}{x}) \notin \Delta$. Hence there is some $\delta(\underset{\sim}{x}) \in \Delta$ such that $\mathcal{A} \models \neg\, \delta(\underset{\sim}{a})$, as required.

COROLLARY 24. Let H be a finitely generated group (ℓ-group) which can be embedded in every existentially complete group (ℓ-group). Then H can be recursively presented with soluble word problem.

Proof. T is finitely (and hence recursively) axiomatizible. Let $c_0,\ldots,c_{n-1}$ be generators of H and let Δ be the set of all formulae of $\mathcal{L}$ which are basic, have free variables only among $x_0,\ldots,x_{n-1}$ and which hold in H at $\underset{\sim}{c}$. Clearly Δ, if not recursive, satisfies the hypohteses of the previous theorem. Hence it is omitted in some finitely generic G. But G is existentially complete in T and H cannot be embedded in G. This contradiction shows that Δ is recursive so H has soluble word problem.

This indicates that telling apart different existentially complete structures is no easy matter. One can't just solicit nice groups (ℓ-groups) to do the trick but only ones that have insoluble word problem. So no recursive algorithm is forthcoming.

Finally we prove Theorem 1.11, that every finitely generic abelian ℓ-group is hyperarchimedean (as might be expected by Theorem 19).

Proof. Recall that G is hyperarchimedean if and only if for every $f, g > 0$, $f \in G(g) \boxplus g^{\perp}$; i.e., there exist f_1, f_2 such that $f_1 \in G(g)$, $f_2 \wedge g = 0$ and $f_1 + f_2 = f$. Now for existentially complete abelian ℓ-groups this is equivalent to the $\forall\exists\forall$ sentence θ: $\forall x_0 > 0\, \forall x_1 > 0\, \exists x_2 > 0\, \exists x_3 > 0(x_2 \wedge x_0 = 0\ \&\ x_3 \in G(x_0)\ \&\ x_2 + x_3 = x_1)$. (Strictly we should replace "$x_3 \in G(x_0)$" by $\forall x_4\, \forall x_5(x_4 + x_5 \neq x_3$ or $x_4 \wedge x_5 \neq 0$ or $x_4 = 0$ or $x_5 = 0$ or $x_4 \wedge x_0 \neq 0)$; i.e., x_3 fails to split over x_0. $(\forall x > 0)\phi(x)$ is short for $\forall x(x > 0 \to \phi(x))$ and $(\exists x > 0)\phi(x)$ is short for $\exists x(x > 0\ \&\ \phi(x))$. If some condition p forced $\neg\theta$, then we may assume $\{c_0 > 0, c_1 > 0\} \subseteq p$ and $p \Vdash \neg \exists x_2 > 0 \neg \exists x_3 > 0$ $(x_2 \wedge c_0 = 0\ \&\ x_3 \in G(c_0)\ \&\ x_2 + x_3 = c_1)$. Now p is consistent so, replacing all constants of C occurring in p by variables to obtain $p(\underset{\sim}{x})$, we have $\exists \underset{\sim}{x} \bigwedge p(\underset{\sim}{x})$ is consistent. By Proposition 22.1, since $\overline{C}(X,\mathbb{R})$ is existentially complete, $\overline{C}(X,\mathbb{R}) \models \exists \underset{\sim}{x} \bigwedge p(\underset{\sim}{x})$. Now $\overline{C}(X,\mathbb{R})$ is hyperarchimedean so for some $n \in \mathbb{Z}^+$, $\overline{C}(X,\mathbb{R}) \models \exists \underset{\sim}{x}(\bigwedge p(\underset{\sim}{x})\ \&\ x_1 \wedge nx_0 = x_1 \wedge (n + 1)x_0)$. Hence $p \cup \{c_1 \wedge nc_0 = c_1 \wedge (n + 1)c_0\}$ is a condition. Let $c_2, c_3 \in C$ not occur in p. Then $q = p \cup \{c_1 \wedge nc_0 = c_1 \wedge (n + 1)c_0, c_1 = c_2 + c_3, c_3 = c_1 \wedge nc_0\}$ is a condition extending p, and so $q \Vdash \neg (c_2 \wedge c_0 = 0\ \&\ c_3 \in G(c_0)\ \&\ c_2 + c_3 = c_1)$. But $c_2 \wedge c_0 = [c_1 - (c_1 \wedge nc_0)] \wedge c_0 = [0 \vee (c_1 - nc_0)] \wedge c_0 =$ $0 \vee [(c_1 - nc_0) \wedge c_0] = 0 \vee ([c_1 \wedge (n+1)c_0] - nc_0) = 0 \vee [(c_1 \wedge nc_0) - nc_0] =$ $0 \vee [(c_1 - nc_0) \wedge 0] = 0$. Hence $r = q \cup \{c_2 \wedge c_0 = 0\}$ is a condition containing q so $r \Vdash \neg (c_2 \wedge c_0 = 0\ \&\ c_3 \in G(c_0)\ \&\ c_2 + c_3 = c_1)$. Since $c_2 \wedge c_0 = 0$, $c_2 + c_3 = c_1 \in r$, $r \Vdash \neg c_3 \in G(c_0)$ (and recall $c_3 = c_1 \wedge nc_0$!); i.e., given $s \supseteq r$, there is $t \supseteq s$ such that $t \Vdash \exists x_4\, \exists x_5(x_4 + x_5 = c_3\ \&$ $x_4 \wedge x_5 = 0\ \&\ x_4 \neq 0\ \&\ x_5 \neq 0\ \&\ x_4 \wedge c_0 = 0)$. So for some $c_4, c_5 \in C$, $\{c_4 + c_5 = c_3, c_4 \wedge c_5 = 0, c_4 \neq 0, c_5 \neq 0, c_4 \wedge c_0 = 0\} \subseteq t$. Now $0 < c_4 = c_4 \wedge c_3 = c_4 \wedge (c_1 \wedge nc_0) \leq c_4 \wedge nc_0 = 0$ since $(c_4 \wedge c_0 = 0) \in t$, a contradiction. Hence no condition forces $\neg\theta$. Thus $\emptyset \Vdash \neg\neg\theta$, as required.

FOOTNOTES

1. "Existentially complete abelian lattice-ordered groups" (section 1 of this article) will appear in *Trans. Amer. Math. Soc.* "Existentially complete lattice-ordered groups" (section 2 of this article) will appear in *Israel J. Math.*

2. A proof, assuming Lemma 1.12, may be found at the end of the Appendix.

3. in *Trans. Amer. Math. Soc.*

4. This is not altogether flippant. The only algebraic sentence which we know of for distinguishing finitely generic division rings from infinitely generic ones (due to Angus Macintyre -- unpublished) is over three times as complicated.

5. Note that $\mathscr{A}$ and $\mathscr{D}$ used in the appendix do not have the same meaning as in section 1. We apologize for using the same symbol for two distinct meanings. It was not our original intention but was forced on us by typographical consideratons.

REFERENCES

1. R. N. Ball, Topological lattice ordered groups, *Pacific J. Math. 83:* 1-26 (1979).

2. G. Cherlin, *Model-Theoretic Algebra--Selected Topics*, Springer Lecture Notes, No. 521, Heidelberg (1976).

3. P. F. Conrad, Regularly ordered groups, *Proc. Amer. Math. Soc. 13:* 726-731 (1962).

4. P. F. Conrad, Epi-archimedean groups, *Czech. Math. J. 24:* 1-27 (1974).

5. P. C. Eklof, Ultraproducts for algebraists, in *Handbook of Mathematical Logic*, ed. K. Jon Barwise, North Holland, Amsterdam (1977).

6. C. D. Fox, Commutators in orderable groups, *Comm. in Algebra 3:* 213-217 (1975).

7. A. M. W. Glass, The word problem for lattice-ordered groups, *Proc. Edinburgh Math. Soc. 19:* 217-221 (1975).

8. A. M. W. Glass, *Ordered Permutation Groups*, Bowling Green State University (1976).

9. Y. Gurevich, Hereditary undecidability of the theory of lattice-ordered abelian groups, in Russian, *Algebra and Logic 6:1:* 45-62 (1967).

10. J. Hirschfeld and W. H. Wheeler, *Forcing, Arithmetic, Division rings*, Springer Lecture Notes, No. 454, Heidelberg (1975).

11. L. Lipschitz and D. Saracino, The model companion of the theory of commutative rings without nilpotent elements, *Proc. Amer. Math. Soc. 38:* 381-387 (1973).

12. A. Macintyre, Omitting quantifier-free types in generic structures, *J. Symbolic Logic 37:* 512-520 (1972).

13. A. Macintyre, On algebraically closed groups, *Annals of Math. 96:* 53-97 (1972).

14. A. Macintyre, Model completeness, in *Handbook of Math. Logic*, ed. K. Jon Barwise, North Holland, Amsterdam (1977).

15. K. R. Pierce, Amalgamations of lattice-ordered groups, *Trans. Amer. Math. Soc. 172:* 244-260 (1972).

16. A. Robison and E. Zakon, Elementary properties of ordered abelian groups, *Trans. Amer. Math. Soc. 96:* 222-236 (1960).

17. W. H. Wheeler, A characterization of companionable, universal theories, *J. Symbolic Logic 43:* 402-429 (1978).

ORDERED ABELIAN GROUPS

Yuri Gurevich

Ben Gurion University of the Negev
*Be'er Sheva, Israel**

This is a propaganda for [1].

In happiest dreams topologists invent invariants characterizing topological spaces modulo homomorphisms. Group theorists would like to learn everything about groups. In practice topologists study special classes of topological spaces and special features of topology. Similarly do group theorists.

What does it mean to learn everything about topology? Is the notion of metric topological? Is the notion of potato topological? I believe the answer to the last question is no. But isn't it natural to try to define precisely the language of topology? Of course, different formalizations are possible, more or less restrictive. We come this way to different specifications of the problem to learn everything about our theory.

Formalization itself does not necessarily imply restriction. The whole mathematics is comfortable in the first order version of ZFC. (Did opponents of the first alphabet argue that sounds of human language can't be listed?) But here we are interested in a formal language L which simplifies the informal theory, so that we have a chance to learn everything about our models which is expressible in L. Of course, such a formalization is interesting only if it mirrors interesting features of the informal theory.

In [1] we study the theory (call it T) of (linearly) ordered abelian groups with quantification over (elements and) convex subgroups. We classify ordered abelian groups modulo indistinguishability in T and prove decidability of T. We hope T is rich enough to be of interest. See decidability results on ℓ-groups in [2].

---

*Paper presented while on leave to Department of Mathematics, Simon Fraser University, Burnaby, British Columbia

REFERENCES

1. Y. Gurevich, Expanded theory of ordered abelian groups. *Annals of Math. Logic, 12:* 193-228 (1977).

2. Y. Gurevich, A contribution to the elementary theory of lattice ordered abelian groups and K-lineals. *Soviet Mathematics*, Doklady *8:* 987-989 (1967).

about the book . . .

Ordered Groups: Proceedings of the Boise State Conference brings together an outstanding collection of original papers which gives the reader an up-to-the-minute view of research in the field of ordered structures. The eminent group of mathematicians whose work is contained in this book present the technical results of their research, and the new uses of ideas and techniques from other areas of mathematics being made in the study of ordered structures.

These chapters focus on a variety of topics that exhibit the diversity and depth of current research in the field. Special emphasis is placed on fundamental discussions of the recent impact of ideas from mathematical logic and model theory on lattice-ordered groups. These chapters also present recent elegant permutation results, an examination of the influence of Cauchy space techniques on completion results, and purely algebraic discussions emphasizing structural questions.

Graduate students and researchers in ordered structures will find this book valuable both as a record of the work of the most eminent mathematicians in this field, and as an exciting source of new problems and ideas for their own research.

about the editors . . .

Jo E. Smith is Assistant Professor of Mathematics, General Motors Institute, Flint, Michigan. She received a B.S. degree (1971) in mathematics and physics, an M.A. degree (1972), and a Ph.D. degree (1976) in mathematics from Bowling Green State University, Bowling Green, Ohio. Dr. Smith has published articles on varieties of lattice-ordered groups and is also interested in graph theory and general relativity.

G. Otis Kenny is an Associate Professor of Mathematics at Boise State University, Idaho. Dr. Kenny has published several articles on lattice-ordered groups and is presently studying computer modeling and simulation. He received his A.B. degree (1970) from Earlham College, Richmond, Indiana; and his M.A. degree (1973), and Ph.D. degree (1975) in mathematics from the University of Kansas, Lawrence.

Richard N. Ball is an Associate Professor of Mathematics at Boise State University, Idaho. He received his B.A. degree (1966) from the University of Colorado, Boulder; and his M.A. degree (1967) and Ph.D. degree (1974) in mathematics from the University of Wisconsin, Madison. Dr. Ball has published several articles on ℓ-group completion theory.

Printed in the United States of America

ISBN: 0–8247–6943–0

marcel dekker, inc./new york · basel